Atheer Barghouthi
A Fully Integrated Phase Locked Loop at 61.44 GHz
for High-Speed Wireless LANs

TUDpress

Atheer Barghouthi

A Fully Integrated Phase Locked Loop at 61.44 GHz for High-Speed Wireless LANs

TUDpress

2013

Die vorliegende Arbeit wurde am 01. Juni 2012 an der Fakultät Elektrotechnik und Informationstechnik der Technischen Universität Dresden als Dissertation eingereicht und am 25. Oktober 2012 erfolgreich verteidigt.

Vorsitzender der Promotionskommission:
Prof. Dr.-Ing. Plettemeier

Gutachter:
Prof. Dr.-Ing.habil Ellinger
Prof. Dr.-Ing.habil Schumacher

Bibliografische Information der Deutschen Nationalbibliothek
Die Deutsche Nationalbibliothek verzeichnet diese Publikation in der Deutschen Nationalbibliografie; detaillierte bibliografische Daten sind im Internet über http://dnb.d-nb.de abrufbar.

Bibliographic information published by the Deutsche Nationalbibliothek
The Deutsche Nationalbibliothek lists this publication in the Deutsche Nationalbibliografie; detailed bibliographic data are available in the Internet at
http://dnb.d-nb.de.

ISBN 978-3-944331-04-1

© 2013 TUDpress
Verlag der Wissenschaften GmbH
Bergstr. 70 | D-01069 Dresden
Tel.: 0351/47 96 97 20 | Fax: 0351/47 96 08 19
http://www.tudpress.de

Alle Rechte vorbehalten. All rights reserved.
Gesetzt vom Autor.
Printed in Germany.

Technische Universität Dresden

A Fully Integrated Phase Locked Loop at 61.44 GHz for High-Speed Wireless LANs

Atheer Barghouthi

von der Fakultät Elektrotechnik und Informationstechnik der Technischen Universität Dresden

zur Erlangung des Akademischen Grades eines

Doktoringenieurs

(Dr.-Ing.)

genehmigte Dissertation

Vorsitzender: Prof. Dr.-Ing. Plettemeier

Gutachter : Prof. Dr.-Ing. habil. Ellinger   Tag der Einreichung: 01.06.2012

Prof. Dr.-Ing. habil. Schumacher   Tag der Verteidigung: 25.10.2012

This work is dedicated to my wife and daughter, Sahar and Sama, to my parents, Sami and Hakima, to my brothers, Oban and Aseel, and to my sister, Bana.

Table of Content

Abstract VI
Chapter 1 Introduction 1
  1.1   60 GHz Wireless Communications Systems 1
  1.2   60 GHz Band, Channelization and Regulations 1
  1.3   High Speed Wireless Communication Requirements 3
      1.3.1   Bandwidth Requirements 3
      1.3.2   Technology Requirements 3
  1.4   EASY-A Project 4
  1.5   Thesis Outline 6
Chapter 2 Wireless System Architectures 8
  2.1   Introduction 8
  2.2   System Specifications 8
      2.2.1   Transmitter Specifications 8
          2.2.1.1 Error Vector Magnitude (EVM) 8
          2.2.1.2 Spectral Mask 9
      2.2.2   Receiver Specifications 10
          2.2.2.1 Sensitivity 10
          2.2.2.2 Dynamic Range 11
  2.3   System Impairments 11
      2.3.1   I/Q Imbalance 11
      2.3.2   DC Offsets and Local Oscillator (LO) Leakage 12
      2.3.3   Phase Noise 12
      2.3.4   Nonlinearity 13
      2.3.5   Flicker Noise 14
  2.4   System Architecture 14
      2.4.1   Transmitters Architectures 14
          2.4.1.1 Superheterodyne Tranmitter 14
          2.4.1.2 Sliding IF 15
          2.4.1.3 Direct Conversion 15
      2.4.2   Receivers Architectures 15
          2.4.2.1 Superheterodyne Receiver 15
          2.4.2.2 Sliding IF 16

2.4.2.3 Direct Conversion 16
2.5 Adopted Architecture 17

Chapter 3 SiGe HBT BiCMOS Technology 19
3.1 Introduction 19
3.2 Technology Figure of Merits (FOMs) 19
3.3 SiGe HBT 20
3.4 SiGe HBT BICMOS 20
3.5 IHP's SiGe HBT BiCMOS Technology 20
  3.5.1 Technology Introduction 20
  3.5.2 Technology Passive Components 21
    3.5.2.1 Resistors 21
    3.5.2.2 MIM Capacitors 21
    3.5.2.3 Transmission Lines and Inductors... 22
    3.5.2.4 Varactors 23

Chapter 4 Charge-Pump-Based PLL System Modeling 24
4.1 Introduction 24
4.2 PLL Non-idealities 24
  4.2.1 Phase Noise 25
  4.2.2 Spur Power 25
  4.2.3 Dynamic Behavior... 26
4.3 PLL Specifications 27
4.4 System Architecture 27
4.5 PLL Analysis Model.. 28
4.6 Verilog-A Modeling 29
4.7 Phase Noise Modeling 29
  4.7.1 Direct Simulation 30
  4.7.2 Time Domain Simulation 30
  4.7.3 Frequency Domain Simulation.. 30
4.8 Simulations 31
4.9 PLL Trade-offs 32

Chapter 5 Voltage Controlled Oscillators.. 34
5.1 Introduction 34
5.2 Oscillator Models 34
  5.2.1 The Positive Feedback Approach 34

  5.2.2 The Negative Resistance Approach 35
 5.3 Varactor Figure of Merits (FOMs) 36
  5.3.1 The Quality Factor (Q-factor) 36
  5.3.2 The Tuning Ratio 37
  5.3.3 The Self Resonance Frequency (SRF) 37
 5.4 Varactor Types 37
  5.4.1 PN-diode Varactors 37
  5.4.2 MOSFET Varactors 38
   5.4.2.1 Inversion Mode MOSFET 38
   5.4.2.2 Accumulation Mode MOSFET 38
 5.5 Varactor Modeling 39
  5.5.1 Simplified Circuit Model 39
  5.5.2 Generalized Circuit Model 40
  5.5.3 Measurements and Testing 42
 5.6 VCO Properties 44
  5.6.1 Tuning Range 44
  5.6.2 Phase Noise 44
   5.6.2.1 Leeson's Model 44
   5.6.2.2 LTV Model 45
  5.6.3 Output Power 47
 5.7 Types of LC VCOs 47
  5.7.1 Cross Coupled VCO 48
  5.7.2 Common Collector Differential Colpitts VCO 52
   5.7.2.1 Design Trade-offs 52
   5.7.2.2 The effect of Parasitics and Load Resistance on the Negative Resistance 54
   5.7.2.3 Components Sizing 56
 5.8 Common Collector Colpitts Quadrature VCO (QVCO) 58

Chapter 6 Frequency Dividers 62
 6.1 Introduction 62
 6.2 Divider Characteristics 62
  6.2.1 Self-Oscillation Frequency 62
  6.2.2 Divider Output Power 62
  6.2.3 Divider Sensitivity Curve 62
 6.3 Divider Types 63

- 6.3.1 Dynamic Dividers 63
- 6.3.2 Static Dividers 64
- 6.4 Static Dividers Latch Architectures 65
  - 6.4.1 Conventional Latch 65
  - 6.4.2 Conventional Latch with Peaking 65
  - 6.4.3 Split-Load Latch 66
  - 6.4.4 Emitter Coupled Logic Latch 67
- 6.5 Simulations and Comparison 68

Chapter 7 Phase Frequency Detectors and Charge Pumps (PFD/CP) 69

- 7.1 Introduction 69
- 7.2 The Conventional PFD 69
- 7.3 PFD/CP Gain and Dead-zone 70
- 7.4 Charge Pumps 71
- 7.5 Charge Pump Architectures 71
  - 7.5.1 Conventional Charge Pump 71
  - 7.5.2 Highly Matched Charge Pump 72
  - 7.5.3 Highly Matched Charge Pump with Reduced Voltage Headroom 73
- 7.6 CP Design Issues 73
  - 7.6.1 Charge Pump Noise 73
  - 7.6.2 Current Sources Voltage Headroom 75
  - 7.6.3 Current Sources Output Impedance 75
  - 7.6.4 Charge Pump Reference Feed-through 75

Chapter 8 Loop Filter 77

- 8.1 Introduction 77
- 8.2 Loop Filter Types 77
- 8.3 Loop Filter Order 79
- 8.4 Integrated versus Non-integrated Loop Filters 79
- 8.5 Capacitor Multiplier for Integrating Loop Filters 79
  - 8.5.1 The Conventional Capacitor Multiplier 79
  - 8.5.2 The Self-biased Capacitor Multiplier 80

Chapter 9 Physical Design, Simulation and Measurement Results 83

- 9.1 Introduction 83
- 9.2 VCO 83
- 9.3 Divide-by-1024 Frequency Divider 85

9.4	Loop Filter	86
9.5	Output Splitter	88
9.6	Integrated PLL	89
9.7	Partly-integrated PLL	91
9.8	System Measurements	94
9.9	Common Collector Colpitts QVCO Simulations and Measurements	97

Chapter 10 Conclusion and Perspective **100**

- 10.1 Introduction 100
- 10.2 Comparison with State-of-the-Art 100
  - 10.2.1 VCO 100
  - 10.2.2 PLL 100
  - 10.2.3 QVCO 101
- 10.3 Theoretical Extensions 101
- 10.4 Outlook 102

Appendix A **103**

A. Verilog-A PLL Codes **103**

- A.1 VCO Verilog-A Code 103
- A.2 Divider Verilog-A Code 104
- A.3 PFD Verilog-A Code 105
- A.4 Charge-pump Verilog-A Code 106

Appendix B **107**

B. HBT Small Signal Model **107**

Appendix C **108**

C. Matlab Codes **108**

- C.1 Settling Time Simulations 108
- C.2 PLL Loop Bandwidth 108

References **110**

Publications **114**

Terms and Abbreviations **115**

List of Figures **117**

List of Tables **120**

Curriculum Vitae **121**

Abstract

The availability of a large unlicensed bandwidth at the 60 GHz band combined with the ever increasing need for higher data rates encouraged the research community to investigate wireless transceivers at this frequency band. The EASY-A project investigates two kinds of applications in the 60 GHz band. The very high rate extended range application (VHR-E), which targets a data rate of up to 4 Gbps within a range of 10 meters, and the ultra-high rate cordless (UHR-C) application, which targets a data rate of up to 10 Gbps within a range of 1 meter. At high frequencies the gain of amplifiers becomes more challenging and more difficult to achieve. In addition, at higher data rates, high order modulation schemes might be necessary, which makes the phase noise specifications of PLLs more difficult to be met. Therefore, advanced technologies which can handle the high speed of operation and can still have the high integration level, low cost advantage are required. SiGe BiCMOS technologies are one of the best options for such applications.

For the implementation of the analog front-end of both applications, local oscillators are indispensable components. These can be implemented using phase locked loops. In this work, the design of the 61.44 GHz PLL for the UHR-C transmitter of the EASY-A project is introduced. Two versions of the PLL were designed: version 1, which is fully integrated, and version 2, which is partly integrated. Version 2 targets higher-end applications, where better phase noise performance is needed. To implement the PLL, a 60 GHz VCO, an output splitter, a divide-by-1024 frequency divider, a phase frequency detector, and a loop filter were designed and combined to achieve the PLL functionality.

In this work, several circuits were designed and measured and some of them were published in leading conference and journal contributions:

- A 60 GHz VCO, which is based on the differential common collector Colpitts VCO architecture was designed and measured. Within a wide tuning range from 54.1 to 62.7 GHz, a phase noise performance between -98 and -90 dBc/Hz was measured at 1 MHz offset. To our knowledge, this is the best phase noise performance of a 60 GHz VCO in silicon yielding such a high tuning bandwidth. At a dc voltage of 2 V and a dc current of 14 mA, the single-ended output power amounts to -5 ± 1 dBm.

- A divide-by-1024 60 GHz frequency divider was designed and measured. The output of the divider is a 0 to 2 V periodic 60 MHz square wave signal. The divider consumes a power of 90 mW using a 3 V supply voltage.

- A 60 GHz QVCO, which is based on the parallel coupling of two differential common collector Colpitts VCOs was designed and measured. At a dc bias voltage of 2 V and current consumption of 32 mA, each output delivers -13 ± 1.5 dBm of power. An excellent phase noise performance in the range of -101 to -96 dBc/Hz at 1 MHz offset was measured in a tuning bandwidth of 2.3 GHz, from 59.7 to 62 GHz. To our knowledge, this is the best phase noise performance reported for 60 GHz QVCO in silicon.

- PLL version 1: a fully integrated 61.44 GHz PLL was designed and measured. The PLL was optimized for a very fast settling time of 4 µs as required by the system specifications. Because the receiver is using a carrier recovery circuit, which can follow the slow changes of the carrier such as phase noise, the sensitivity of the bit error rate to phase noise at the receiver end is very low. As a result, to achieve the required dynamic behavior, the phase noise performance could be sacrificed and the loop bandwidth was increased until the needed settling time was achieved, while taking the suppression of the reference spurs into consideration. Capacitor multiplication was used to enable the integration of the loop filter on chip and the effect of the capacitor multiplier on the total PLL phase noise performance was quantified and evaluated. In addition, a very close matching between the measured and simulated phase noise of the system was achieved. The PLL consumes a power of 200 mW from 2 V and 3 V supply voltages, while delivering a differential output power of -7 dBm, sufficient to drive the following I/Q modulator without additional amplification. The measured phase noise at 1 MHz offset frequency is -71 dBc/Hz.

- PLL version 2: a partly integrated 61.44 GHz PLL for higher-end applications was also designed. The PLL was optimized for a better phase noise performance. The loop filter was implemented off-chip in order to enable the loop-bandwidth adjustment. The loop bandwidth was reduced from 1.6 MHz to 400 kHz. As a result, the phase noise performance was improved by 12 dB, from -71 dBc/Hz to -83 dBc/Hz at 1 MHz frequency offset. The measured power ratio between the carrier and the spurs was improved by 12 dB from -28 to -40 dBc. The PLL consumes a power of 200 mW from 2 V and 3 V supply voltages, while delivering a differential output power of -9 dBm, sufficient to drive the following I/Q modulator without additional amplification.

In addition to the practical achievements, theoretical extensions, in the field of oscillators, to the available literature have been achieved.

- The effect of the HBT transistor parasitics on the negative resistance of Colpitts oscillators was analyzed and the negative resistance equation, which is available in literature, was extended. In addition, a design methodology for choosing the load impedance of Colpitts oscillators was explained.

- The frequency limits of the operation of HBT cross coupled oscillators at high frequencies were studied. An extended expression of the negative conductance of HBT cross coupled oscillators was derived. Furthermore, an expression for the frequency, at which the conductance of the oscillator changes its value from negative to positive, was also derived.

VIII

Chapter 1 Introduction

1.1 60 GHz Wireless Communications Systems

The demands for high-speed wireless data communications, such as, the wireless transmission of huge data files and real-time video, are markedly increasing. The assignment of a large unlicensed bandwidth (5 to 7 GHz) of internationally available spectrum in the 60 GHz frequency-band has the ability to accommodate up to 10 Gbps, which makes the 60 GHz band a suitable candidate to satisfy the demands for high data-rate wireless communications. Fig 1.1 shows the trend of data-rates in wireless communication systems from the year 1995 to 2015. Unfortunately, due to the wireless channel propagation characteristics at frequencies around 60 GHz, where the attenuation of signals is high, 60 GHz applications are limited to short-ranges. In addition, 60 GHz implementations must overcome several challenges, such as, the limited gain in amplifiers and the excessive phase noise in oscillators.

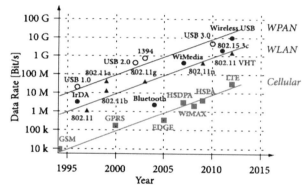

Fig 1.1 Trend of data rates in wireless communications [Pet08]

1.2 60 GHz Band, Channelization and Regulations

The bandwidth regulations of the 60 GHz band depend on the location, as shown in Fig Fig 1.2. The target transceiver is intended to comply with most of the worldwide regulations of the 60 GHz frequency-band. Therefore, it should follow the IEEE 802.15.3c / ECMA-387 channelization, which is depicted in Fig 1.3. Bonding of two channels with a total bandwidth of 4.32 GHz is required to achieve the target 10 Gbps data-rate. Channels 2 and 3 in Fig1.3 are used, which leads to a carrier frequency of 61.56 GHz. It should be noticed that, this channel bonding scheme will exclude Japan and Australia from the transceiver applications. For Germany, the RTTT band overlaps with channel 3. If this implies any restrictions on using channel 3 in Germany, channels 1 and 2 could be used instead.

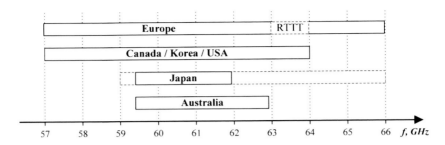

Fig 1.2 60 GHz frequency band [ECMA]

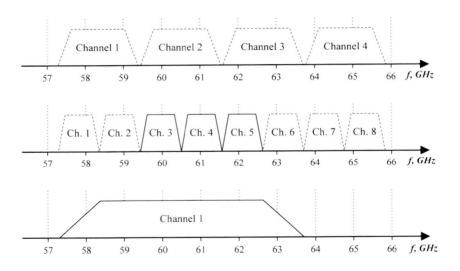

Fig 1.3 IEEE 802.15.3c / ECMA-387 channel plan [ECMA]

In addition to the channel regulations, the 60 GHz band regulations also define the limit of the maximum allowed transmit output power in terms of maximum transmitter power into the antenna and peak effective isotropic radiated power (EIRP). The power regulations are summarized in Fig 1.4. A maximum of 10 dBm transmit power is allowed in most countries except North America.

Chapter 1. Introduction 3

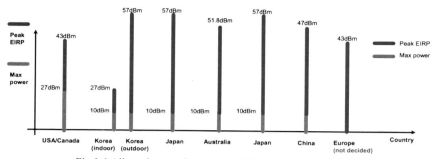

Fig 1.4 Allowed transmit power in different countries [ECMA]

1.3 High Speed Wireless Communication Requirements

In this section, both the bandwidth and technology requirements for achieving high data rates using the 60 GHz band are explained.

1.3.1 Bandwidth Requirements

In order to transfer higher data rates, the symbol rate R_s, as defined in (1.1), should be increased.

$$R_s = \frac{1}{T_s} \qquad (1.1)$$

where T_s is the symbol duration. This can be done by using a larger bandwidth. In addition, using a higher order modulation schemes, such as quadrature modulation, increases the bit-rate, which can be transferred in the same bandwidth. The bit-rate can be calculated from the symbol rate when the spectral efficiency (n) of the modulation scheme is taken into account. For example, QPSK has a spectral efficiency of 2 bit/s/Hz. As a result, the bit-rate R_b bit/s can be calculated as in (1.2).

$$R_b = nR_s \qquad (1.2)$$

Sending baseband data with a symbol rate of R_s symbol/s requires a baseband bandwidth of $R_s/2$ Hz. This bandwidth corresponds to R_s Hz when the data is up-converted to become a pass-band signal. As a result, using QPSK as a modulation scheme, gives a data rate of $2R_s$ bit/s in a pass-band bandwidth of R_s. The effective bandwidth of bonding two channels in the 60 GHz band is 3,456 GHz. Hence, one spatial stream using QPSK provides a data-rate of 6,912 Gbps. As a result, two spatial streams result in a data-rate of 13.824 Gbps. To calculate the effective data-rate, the code rate (commonly used 3/4 or 5/6) and preambles should be taken into account. As a result, the effective data rate will be approximately 10 Gbps.

1.3.2 Technology Requirements

In order to enable the transmission of high data-rates up to 10 Gbps, in addition to the bandwidth requirements of the system, a high speed semiconductor technology for the implementation of the analog front-end should also be available. Otherwise, using a slower technology will require the designer to develop unique architectures, which is a very challenging task. In addition to the speed requirement, components models at 60 GHz should also be as accurate as possible. This allows the designer to design efficiently and accurately.

It is well-known that the III-V semiconductors technologies, such as, GaAs and InP, allows the design of high performance, high-speed analog circuits. Unfortunately, their high costs and low integration levels eliminate them from being an option for this project. Advanced CMOS technologies are another option, which is low cost and has high integration levels. However, the availability of the SiGe heterojunction bipolar transistors (HBT) BiCMOS technology from IHP made it the suitable choice for this project. This technology has a transit frequency f_t of 180 GHz and a maximum frequency of oscillation f_{max} of 200 GHz, which allows the design of circuits in the 60 GHz domain. More details about the technology are given in Chapter 3.

1.4 EASY-A Project

EASY-A stands for "Enablers for Ambient Systems Part-A". The EASY-A project targets two types of applications: the very high rate extended range application (VHR-E), where a data rate of 4 Gbps is transmitted over a distance of 10 meters, and the ultra-high rate cordless applications (UHR-C). In this thesis, only the UHR-C type is considered. The UHR-C physical layer (PHY) is mainly intended for kiosk scenario applications, but can also be applied for computer tomography (CT) applications. Fig 1.5 shows a picture that clarifies the kiosk scenario. A well-known application of the kiosk scenario is the download of very large amounts of bulk data from a machine. The user is situated in front of the machine, at a distance of less than 1 meter. A line of sight (LOS) communication link between the transmitter and receiver can be assumed, because the amount of scatterers in the direct vicinity is limited. The downloaded data is stored in the user terminal. The data has to be transferred in an acceptable amount of time, typically less than 10 s. The user terminal is a battery powered handheld device, with corresponding requirements on the form factor and power consumption.

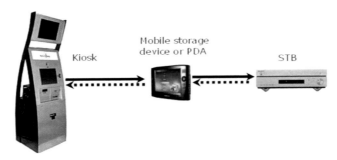

Fig 1.5 Kiosk scenario file download (STB is a server) [Sad09]

Both constraints are less severe for the transmitting device, which have access to the main supply.

The above description implies the following characteristics and requirements for the UHR-C PHY, [Gra08]:

- Data rate of up to 10 Gbps
- Transmission range of up to 1 meter
- AWGN-like physical transmission channel (excluding antennas and analog transceiver front-ends). The spatial diversity is limited to polarization diversity.
- Time division duplexing for separation of uplink and downlink, due to the small distances and highly asymmetric traffic.
- Long packets (on the order of several kbytes), to ensure low PHY overhead for acquisition, synchronization, signaling, etc.
- Sufficiently low bit error rate for "error-free" transmission on top of the transport layer, i.e., in the range of 10^{-12} including appropriate retransmission protocols. The frame error rate must be low enough to ensure acceptable performance of the transport/network layer.
- Very low power consumption of the user terminal.

The UHR-C PHY design can be done in two different approaches, [Kro11]:

1. Analog oriented solution: in this approach, most of the baseband processing is shifted to the analog domain. This includes pulse shaping and generation of the modulated signals at the transmitter, as well as synchronization and demodulation at the receiver. This approach is expected to enable a very low power consumption of the user terminal.

2. Digital oriented solution: this solution uses DACs and ADCs of appropriate resolution at the transmitter and receiver. For the generation of the signals at the transmitter and acquisition, synchronization and demodulation at the receiver, methods well-known from wireless LAN systems are used. This approach includes frequency domain channel equalization at the receiver but comes with the drawback of higher expected power consumption of the user terminal.

The analog oriented solution was adopted for the UHR-C transceiver, as shown in Fig 1.6. This choice was driven by the low power consumption of the architecture. In addition, state-of-the-art ADCs and DACs would not have allowed for an implementation of the digital oriented solution. The parameters of the UHR-C PHY are given in Table 1.1, [Kro11]. Regarding the choice of the modulation technique, orthogonal frequency division multiplexing (OFDM), single carrier (SC) modulation, and dual carrier modulation are the available options for the UHR-C PHY. Because of its high robustness towards analog frontend imperfections, SC modulation has finally been selected. Besides, OFDM would have required substantial digital signal processing (e.g. DFT and IDFT), which would not have complied with the analog oriented approach. Furthermore, the dual carrier modulation would have come with the drawback of a more complex analog transceiver frontend. To allow for an analog carrier recovery at the receiver, which may involve a phase ambiguity (of e.g. 90°), differential de-/modulation techniques are required. As a result, differential QPSK modulation scheme was chosen, which also has the benefit of allowing one-bit ADCs and DACs to be used.

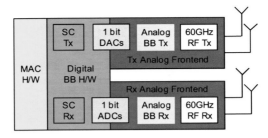

Fig 1.6 UHR-C transceiver architecture [Kro11]

Table 1.1 UHR-C SC-QPSK PHY [Kro11]

Channel plan		IEEE 802.15.3c/ECMA TC48 (channels 2+3)
Antenna setup		2x2 MIMO
BB processing approach		Analog
Modulation scheme		SC
Modulation alphabet		(D)QPSK
Modulation raw spectral efficiency	bpcu	2
Number of used channels		1
Number of spatial streams		2
Code rate		3/4
SC symbol rate	Gsymbol/s	3,456
SC symbol spacing	ns	0,2894
Data SC block length	SC symbols	256000
	ns	74074,07
Total data PHY packet (SC block) length	SC symbols	256000
	ns	74074,07
Preamble / control information SC block length	SC symbols	2048
Preamble length	SC blocks	3
Control information length	SC blocks	1
Total frame length	SC symbols	264192
	us	76,44
Frame payload (per channel and spatial stream)	kByte	48
Frame overhead	%	3,1
total time domain overhead	%	3,1
Effective data rate	Gbps	10,05

1.5 Thesis Outline

In this thesis, the phase locked loop (PLL) design for the transmitter will be covered. In Chapter 2, several transmitter and receiver architectures are explained. The advantages and disadvantages of each are pointed out, which allows the final choice of the analog front-end architecture. In addition, Chapter 2 also includes explanation of the analog front-end impairments. In Chapter 3, the technology which was used for the design is introduced. Chapter 4 presents the PLL verilog-A modeling, which helps in predicting the PLL dynamic behavior. Moreover, the chapter introduces phase noise simulations in the frequency domain, which helps in predicting the phase noise performance of the PLL. In Chapter 5, different 60 GHz VCO architectures and design methodologies are introduced. In Chapter 6, a comparison between different latch topologies is done, which allows the choice of the most suitable topology for the frequency divider. In addition, the design methodology of the 1024-frequency-divider is explained. Chapter 7 deals with the phase frequency detectors. In Chapter 8, the design of the loop filter is introduced. Chapter 9 summarizes the results and achievements of the PLL and the UHR-C transceiver. Finally, in Chapter 10, the thesis is concluded and future work is proposed.

Chapter 2 Wireless System Architectures

2.1 Introduction

A starting point for designing an RF system is to consider the architecture of the analog front-end (AFE), because this choice impacts the system in different aspects. Several criteria exist to help the designer in employing specific AFE architecture, such as, cost, complexity, power consumption, availability of analog-to-digital and digital-to-analog converters, the risk, and integration level, etc. In addition, the interfaces to the channel and to the digital baseband have also to be taken into account. This chapter begins by introducing the system specifications related to the design of the AFE. After that, the most famous architectures are presented. Finally, the chapter ends with introducing the adopted AFE architecture for the UHR-C scenario.

2.2 System Specifications

In order for the system to function properly and not to impact other wireless systems, system specifications for both the transmitter and receiver should be define

2.2.1 Transmitter Specifications

Two important specifications for the transmitter have to be considered: the error vector magnitude and the spectral mask.

2.2.1.1 Error Vector Magnitude (EVM)

As was mentioned in Chapter 1, to increase the data rate of the system in a specified bandwidth, quadrature modulation is used. Ideally, the transmitter should transmit specific constellation points to the receiver. Unfortunately, practically this is not the case. Due to non-idealities in the transmitter design, such as the I/Q imbalance, nonlinearity, bandwidth limitations, and the phase noise of the PLL, the transmitted constellation diagram is degraded and the constellation points are no longer points but clouds or balls around the ideal constellation point. A measure which characterizes this effect is the EVM. The EVM is a single scalar number that gives an indication of the modulation quality of the signal. To calculate the EVM, a comparison between the transmitted symbols and the ideal ones is needed. A sample of M transmitted symbols is taken $T(i)$ and compared to the ideal symbols $I(i)$. Fig 2.1 shows a pictorial description of the EVM. The EVM can be calculated as, [Beh07]:

$$EVM = 20\log\left(\sqrt{\frac{\sum_{i=1}^{M}|T(i)-I(i)|^2}{\sum_{i=1}^{M}|I(i)|^2}}\right) \tag{2.1}$$

Chapter 2. Wireless System Architectures

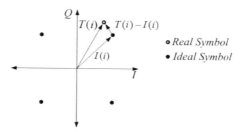

Fig 2.1 EVM calculation

Typical value for the EVM for the WLAN 801.11a and g standards at a data-rate of 54 Mbps is a maximum of -25 dB. The ECMA-387 standard limits the EVM for two bonded channels to a maximum of -9.6 dB.

2.2.1.2 Spectral Mask

A drawback of wireless communication systems in comparison to wired systems is the shared medium of communication between different systems and users. This necessitates the definition of a spectral mask. The spectral mask is the power contained in a specified frequency bandwidth at certain offsets, relative to the total carrier power. By defining the spectral mask, the effect of one wireless system to others operating at frequencies, which is in close vicinity of the system's frequency-band, is minimized.

A linear system generates the same frequency contents as its input. Unfortunately, there are no transmitters which are perfectly linear. Therefore, there is always at least weak nonlinearity in the system. Nonlinear systems cause inter-modulation distortion IMD and harmonic distortion HD to appear at the output of the system. The phenomenon is called spectral re-growth. Spectral re-growth depends on the modulated signal, which can have a constant-envelope, such as, FSK signals, or a non-constant-envelope, such as, QAM signals. The latter is more critical, because it is more susceptible to spectral re-growth. It is worth mentioning that, the most critical component regarding the nonlinearities in the system is the power amplifier. By choosing a constant-envelope modulation scheme the linearity requirements of a power amplifier can be relaxed. Unfortunately, constant-envelope modulation schemes have smaller spectral efficiency, which limits the obtainable data-rate in a specific bandwidth. Therefore, in high data rate applications, non-constant envelope modulation schemes are often used, which makes the linearity of a power amplifier of high concern.

The spectrum of the transmitted signal depends on the spectrum of the modulated signal, the filtering characteristics of the transmitter, and the transmitter nonlinearities. In order to satisfy the specified spectral mask, a pulse shaping filter is normally included at the beginning of the transmitting chain, where the original symbols are filtered to limit their bandwidth. This filtering causes inter-symbol interference, which could degrade the system performance if not carefully considered. The 60 GHz spectral mask, which is defined by the ECMA standard, is shown in Fig 2.2. Table 2.1 specifies the cut-off frequencies of the spectral mask with respect to the channel bonding schemes of the standard.

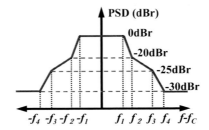

Fig 2.2 Transmitter spectrum mask [ECMA]

Table 2.1 Cut-off frequencies of the spectrum mask [ECMA]

Channel Bonding	Frequency (GHz)			
	f_1	f_2	f_3	f_4
Single Channel	1.05	1.08	1.5	2
Two Bonded Channels	2.1	2.16	3	4
Three Bonded Channels	3.15	3.24	4.5	6
Four Bonded Channels	4.2	4.32	6	8

2.2.2 Receiver Specifications

2.2.2.1 Sensitivity

Receiver sensitivity is the most important parameter in the design of a receiver. It determines the minimum signal power that can be received. The noise figure NF of a receiver is defined as the loss of the signal to noise ratio as the signal passes from the input of the receiver to its output. The NF is calculated as:

$$NF = 10\log\left(\frac{SNR_{in}}{SNR_{out}}\right) \quad (2.2)$$

where SNR_{in} is the signal to noise ratio at the input of the receiver and SNR_{out} is the signal to noise ratio at the output of the receiver. The receiver sensitivity S is related to the NF and is defined as:

$$S = NF + 10\log(kT) + 10\log(BW) + 10\log(SNR_{out}) \quad (2.3)$$

where k is Boltzmann's constant, T is the absolute temperature, and BW is the system bandwidth. The summation of the first three terms is normally called the "noise floor". It can be noticed that the sensitivity of a receiver directly depends on the system bandwidth which is determined by the target data rate.

2.2.2.2 Dynamic Range

In addition to the minimum receivable signal at the input of the receiver, the dynamic range takes into account the maximum tolerable input power at the receiver input. As a result, the dynamic range is defined as the difference in dB between the maximum tolerable signal power at the input of the receiver and the minimum receivable signal. The maximum tolerable signal can be defined as the maximum input power in a two tone test at which the intermodulation products have the same power as the noise floor. Other definitions for the maximum tolerable signal can also be found.

2.3 System Impairments

In this section a short description of the analog front-end impairments that limit the performance of both the receiver and transmitter is included. Such impairments include, I/Q imbalance, phase noise, nonlinearities, etc.

2.3.1 I/Q Imbalance

In modern communication systems, where high data-rates are required, quadrature modulation schemes are applied. To achieve that, quadrature signals have to be generated. The accuracy of the quadrature signals plays an important role in the performance of a communication system. As a result, it is necessary to analyze their impact on the system. Assuming an amplitude imbalance of ε and a phase imbalance of θ, the quadrature signals can be described as, [Raz98]:

$$X_{LO,I(t)}(t) = 2(1+\frac{\varepsilon}{2})\cos(\omega_0 t + \frac{\theta}{2}) \tag{2.4}$$

$$X_{LO,J(t)}(t) = 2(1-\frac{\varepsilon}{2})\sin(\omega_0 t - \frac{\theta}{2}) \tag{2.5}$$

The received signal can be described as:

$$X_{in}(t) = a\cos(\omega_0 t) - b\sin(\omega_0 t) \tag{2.6}$$

After down conversion, the base band signals can be calculated as:

$$X_{BB,I(t)}(t) = a(1+\frac{\varepsilon}{2})\cos(\frac{\theta}{2}) - b(1+\frac{\varepsilon}{2})\sin(\frac{\theta}{2}) \tag{2.7}$$

$$X_{BB,Q(t)}(t) = a(1-\frac{\varepsilon}{2})\cos(\frac{\theta}{2}) + b(1-\frac{\varepsilon}{2})\sin(\frac{\theta}{2}) \tag{2.8}$$

As a result, the amplitude and phase mismatches in the carrier change the constellation of the received symbols at the output of the quadrature demodulator. This effect is shown in Fig 2.3. It can be understood from (2.7) and (2.8) when $\theta=0$ or $\varepsilon=0$ is substituted.

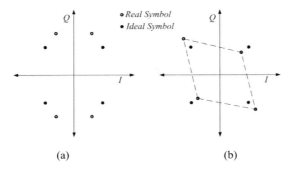

Fig 2.3 I/Q mismatches [Raz98] (a) amplitude mismatch ($\theta=0$)

(b) phase mismatch ($\varepsilon=0$)

2.3.2 DC Offsets and Local Oscillator (LO) Leakage

The main cause of DC offsets in a receiver is self-mixing. If the isolation between the LO port and the RF port in a mixer is poor, three ways can cause self-mixing:

- The LO signal couples to the RF port and mixes with itself.
- The LO signal couples to the low noise amplifier input amplified and then mixes with itself.
- An interferer received by the antenna can couple to the LO port and mixes with itself.

In addition to self-mixing, even order nonlinearities and mismatches in the circuit can worsen the DC offset problem. DC offsets are critical, because they may corrupt the information or saturate the baseband components, hence, degrading the performance of the receiver.

DC offsets can be eliminated by using two approaches: first, a high pass filter can be carefully applied. The cut-off frequency of the filter is a critical parameter in this case, because it controls the information loss at low frequencies and also the transient response of the loop. As a result, a compromise between using a high cut-off or low cut-off frequency has to be used. Second, the DC offsets can be corrected by sampling the DC offsets under low gain settings of the receiver to avoid saturation. Then, digital to analog converters can be used to adjust the DCs in the receiver chain. It is worth mentioning that, the DC offset problem is more severe at high frequencies, due to the reduced isolation. Another issue which becomes more critical at higher frequencies is the LO radiation. If the isolation between the LO and RF port is poor and the reverse gain of the LNA is not too low, the LO signal could couple to the RF port and passe through the LNA to be finally radiated by the antenna.

2.3.3 Phase Noise

Because practically a PLL is not an ideal sinusoidal signal due to phase noise. The effect of phase noise on transmitters and receivers should be studied. Fig 2.4 shows the simulated effect of the transmitter PLL phase noise on the constellation diagram of the transmitted symbols. In Fig 2.4, only phase noise was considered and all other impairments were neglected.

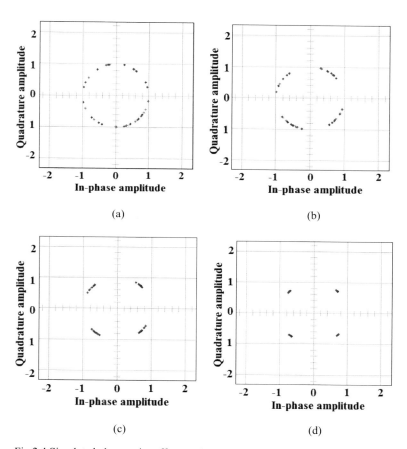

Fig 2.4 Simulated phase noise effect on the constellation diagram of an ideal communication link for different values at 1 MHz offset frequency (a) -71 dBc/Hz (b) -83 dBc/Hz (c) -90 dBc/Hz (d) -100 dBc/Hz

In receiver architectures which have the ability to do carrier recovery in the analog domain, phase noise performance is not of high concern, because the carrier recovery circuit can follow the slow changes in the carrier caused by the phase noise. But, for receiver architectures which uses digital demodulation (with no carrier recovery), phase noise is of high concern and has to be well optimized, [Rei05] and [Ulu10].

2.3.4 Nonlinearity

Another set of receiver impairments is related to the nonlinearities of the receiver. An ideal receiver would faithfully amplify, frequency down convert, and filter the desired signal without creating any nonlinear components at the output. In reality, nonlinearities do exist in a practical receiver and as a result distortion components are generated. The impact of these

14 Chapter 2. Wireless System Architectures

nonlinear components is that they impact the EVM (and therefore quality) of the received signal and increase the BER in the receiver.

2.3.5 Flicker Noise

The effect of flicker noise is critical at the input of the baseband, specially, when the gain of the RF stage is not very high. In (2.9), the power spectrum of flicker noise $S_{flicker}$ for MOSFETs is given [Raz98], which shows strong dependence of flicker noise on the area of the transistor. As a result, when the baseband circuits are designed using MOSFET transistors, the baseband section has to be carefully designed using large transistors to minimize flicker noise effects.

$$S_{flicker} = \frac{K}{A \times f} \quad (2.9)$$

where K is a technology dependent constant, A is the area of the transistor, and f is the frequency. Sometimes, if the data is encoded to have low information content near 0 Hz, high pass filters can be used to attenuate the effects of flicker noise.

2.4 System Architecture

In this section, the most famous transmitter and receiver architectures are introduced. In addition, the Pros and Cons of the architectures are discussed.

2.4.1 Transmitters Architectures

2.4.1.1 Superheterodyne Tranmitter

The superheterodyne transmitter architecture is shown in Fig 2.5. In the superheterodyne transmitter, the data is up-converted to the RF domain in two steps. First, it is up-converted to the IF frequency and then to the RF. Low pass filters at the beginning of the transmitting chain are necessary in order for the transmitter to satisfy the spectral mask requirements. In order to eliminate the image frequencies after mixing, each up-conversion is followed by a band-pass filter.

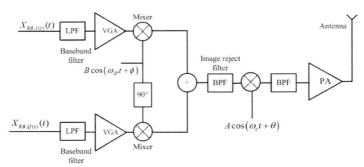

Fig 2.5 Superheterodyne transmitter

2.4.1.2 Sliding IF

Sliding IF is another variant of the superheterodyne transmitter, in which, the low frequency IF signal is generated by dividing the high frequency RF LO signal by an integer. As a result, the IF is not fixed anymore, but sliding. By doing so, the architecture requires only one LO PLL. In addition, image reject mixers could be used to eliminate the need for filters in the architecture [Beh07].

2.4.1.3 Direct Conversion

The direct conversion transmitter architecture is shown in Fig 2.6. In the direct conversion transmitter, the data is directly up-converted to the RF domain without using an IF step. Similar to the superheterodyne case, low pass filters at the beginning of the transmitting chain are necessary in order for the transmitter to satisfy the spectral mask requirements. One problem in this architecture that should be eliminated is the VCO pulling, [Beh07]. Even though, this architecture suffers from VCO pulling, it is the better choice in terms of cost, complexity, power consumption, and integration ability.

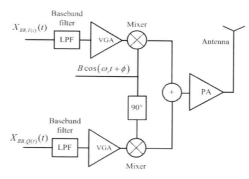

Fig 2.6 Direct conversion transmitter

2.4.2 Receivers Architectures

2.4.2.1 Superheterodyne Receiver

Fig 2.7 shows the architecture of a superheterodyne receiver. In the receiver, the antenna receives wireless signals, which consist of the desired signal together with interferers, depending on its characteristics. The received signal is passed through a band-pass filter which selects the desirable band of communication and attenuates the interferers outside this band, which helps in avoiding desensitizing the receiver. After selecting the desired band, the signal is amplified using a low noise amplifier (LNA) and passed to an image reject filter. The image reject filter filters any signals or noise at the image frequency. The signal is then down-converted into the intermediate frequency (IF) using an RF mixer. The PLL frequency is tuned to transform the desired channel into the IF frequency, which allows a fixed IF band pass filter (BPF) to be used for channel selection. As a result of filtering, the linearity requirements of the blocks following this filter are relaxed and variable gain amplifier can be used to adjust the level of the signal. A second set of mixers is then used to convert the signal into its in-phase and quadrature components by using an LO that has quadrature outputs. At the end of the chain, the baseband signal is passed through an analog-to-digital converter, which allows further processing in the digital baseband.

Some important issues which need to be considered in the design of such a receiver are:

- The band-select filter can be removed if the environment is relatively interference free, or if the LNA has sufficient filtering characteristics and good linearity. By omitting this filter, the *NF* of the receiver is improved by around 1 dB.
- The channel select filter should have high quality factor, which makes the integration of this filter on chip not possible.
- The IF of the system could be chosen to be high or low. A high IF relaxes image rejection but makes the task of channel selection more challenging, whereas, a low IF relaxes the channel selection but makes the band and image rejection more challenging. The IF should be chosen to be larger than half of the baseband bandwidth in order to avoid aliasing at the baseband.

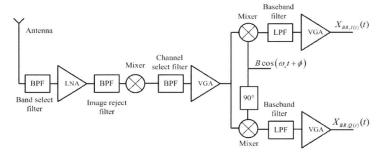

Fig 2.7 Superheterodyne receiver

2.4.2.2 Sliding IF

Sliding IF is another variant of the superheterodyne receiver, in which, the IF LO is generated by dividing the RF LO by an integer. As a result, the IF is not fixed anymore, but sliding. By doing so, the architecture requires only one LO. But the channel filtering in the baseband becomes more complicated.

2.4.2.3 Direct Conversion

In the direct conversion receiver, the RF received signal is down-converted directly to the baseband, as shown in Fig 2.8. The signal is received by the antenna and passed to a band-select filter. The signal is then amplified by an LNA. After that, another set of mixers is used to convert the signal into quadrature baseband signals by using a PLL with the same frequency as the RF carrier. Because the direct conversion architecture does not suffer from the image frequency problem, the number of filters in the architecture is reduced, which makes it more suitable to be integrated than the superheterodyne receiver. In addition, channel selection is done in the baseband with low pass filter, which is easily implemented in comparison to high Q band pass filters. Moreover, the number of components in the receiver is low, which enables a low power consumption solution and allows low cost solution. Unfortunately, many drawbacks exist in this architecture that makes designers avoid adopting it. First, due to using a high frequency quadrature down converter, the I/Q imbalance is worse in this architecture than the superheterodyne receiver. Second, the down conversion from RF directly to DC creates the problem of DC offsets. This is mainly because the coupling and self-mixing at high frequencies is stronger than at lower frequencies. Third, due to the low gain preceding the baseband in this architecture, flicker noise plays important role in the performance of the receiver and should be taken into consideration. Finally, even-order nonlinearities have strong effect on this receiver.

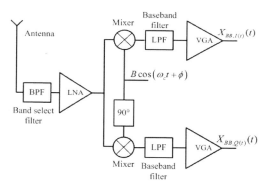

Fig 2.8 Direct conversion receiver

2.5 Adopted Architecture

As was mentioned in Chapter 1, transmitting using the analog approach which does not use DACs is preferred to the digital approach, as a result, on the transmitter side, the direct conversion architecture is the best choice in terms of cost, complexity, power consumption, and integration ability. Therefore, the transmitter architecture shown in Fig 2.9 was chosen for the EASY-A Project.

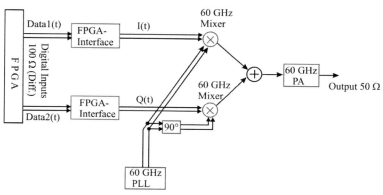

Fig 2.9 Transmitter Architecture

Using QPSK as a modulation scheme makes the receiver architecture which is shown in Fig 2.10, very suitable choice, [Gan12]. By choosing the analog approach, the analog front-end should include as many functions as possible. By doing so, the requirements from the digital domain (i.e. high speed ADCs) are reduced and the realization of the UHR-C receiver using 1 bit DACs (limiting amplifiers) in a compact way and with low power consumption becomes possible. On the other hand, for analog reception, the modulation scheme needs to be kept simple in order to allow synchronization in the analog domain. The proposed way of realizing demodulation and synchronization in the analog domain uses frequency quadrupling for recovering the carrier. A QPSK modulated signal can be represented as:

$$S_{QPSK}(t) = A_c(t)\cos\left(\omega_c t + \frac{\pi}{4}(2m-1) + \phi_n\right) \quad (2.9)$$

where $m \in (1,2,3,4)$ and ϕ_n is an unknown phase. If this signal is passed through a non-linearity, the component at the quadruple frequency would be:

$$S^4_{QPSK}(t) = A_c^4(t)\cos(4\omega_c t + \pi(2m-1) + 4\phi_n) = -A_c^4(t)\cos(4\omega_c t + 4\phi_n) \quad (2.10)$$

The modulated phase reduces to multiples of 2π, and the carrier at the quadruple input frequency remains. Afterwards, the recovered carrier can be used for demodulation as it includes four times the unknown phase. Even though, it is possible to benefit from this method for higher order QAM signals, the target modulation scheme is QPSK, which presents a good trade-off between the spectral efficiency and the required receiver performance. The recovered carrier is directly used for demodulation. The delay experienced in the carrier recovery branch needs to be compensated by time delay elements or a phase shifter network. If the delay is perfectly compensated, the demodulation would be insensitive to phase noise. The intermediate frequency was chosen to be 5 GHz.

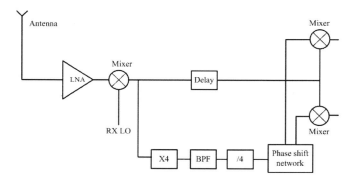

Fig 2.10 Adopted receiver architecture [Gan12]

Chapter 3 SiGe HBT BiCMOS Technology

3.1 Introduction

As was previously mentioned, the continuous demand for higher data rates in wireless communications triggers the need for larger bandwidths. Because more bandwidth is normally available at higher frequencies, high speed wireless communications tend to use higher carrier frequencies. In order to design circuits, such as, power amplifiers, modulators, and oscillators at such high frequencies, the technology used for the design should be capable to handle such high speeds. In addition, the benefits of the economy of scale of silicon technologies, which enables the production of low-cost, very highly integrated, powerful ICs, each containing millions of transistors, have to be maintained.

Because SiGe HBT BiCMOS technology combines both qualities, the high speed and the high integration capability, it became very popular in the past decade for designing high frequency circuits and products. Fig 3.1 shows a cross section of an example, four metal layers, SiGe HBT BiCMOS technology, where NMOS and PMOS transistors are integrated together with HBTs.

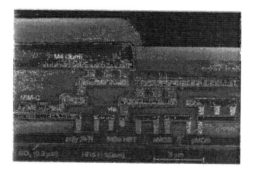

Fig 3.1 Example of a SiGe HBT BiCMOS technology cross section [Cre08]

3.2 Technology Figure of Merits (FOMs)

In order to allow the comparison of two technologies in their high-frequency performance, several FOMs, which describe how fast a transistor can work, are defined:

- The transit frequency f_t: is defined as the transition frequency, at which the common emitter small signal current gain drops to unity for a short circuit as the load. It can be calculated as:

$$f_t \cong \frac{g_m}{2\pi(C_{be} + C_{bc})} \quad (3.1)$$

- The maximum frequency of oscillation f_{max}: is the frequency at which the maximum available power gain $G_{A,max}$ is equal to unity. It is calculated as:

$$f_{max} \cong \sqrt{\frac{f_t}{8\pi C_{bc} R_b}} \qquad (3.2)$$

3.3 SiGe HBT

As stated in [Cre08], even though, the fabrication of silicon wafers and circuits is very attractive, the electrical properties of silicon are not the best when compared to other semiconductors, such as, the III-V semiconductor materials. The electrons mobility in silicon is comparatively small compared to the III–V semiconductors, and the maximum velocity that they can attain under high electric fields is limited to about 10^7 cm/s under normal conditions, which is relatively slow. Because the speed of the device depends mainly on how fast the carrier can be transported under the effect of electric fields, silicon NMOS devices are normally slower than the III-V semiconductors' devices based on n-doping. Even though, the III-V semiconductors devices are faster, III-V semiconductors, such as, GaAs and InP, have lower levels of integration (no complementary circuits can be build using them due to the low mobility of holes and therefore, high static power consumption compared to CMOS circuits), more difficult fabrication procedures, lower yield, and ultimately higher cost, than silicon. As a result, silicon is still the desired semiconductor for large scale applications. In order to enhance the mobility of the carriers in silicon, band-gap engineering principles are used. By including germanium in the lattice structure of silicon, strain occurs, which improves the mobility of carriers, [Cre08].

The principal difference between the BJT and HBT is in the use of differing semiconductor materials for the emitter and base regions, creating a hetero-junction. The effect is to limit the injection of holes from the base into the emitter region, since the potential barrier in the valence band is higher than in the conduction band. Unlike BJT technology, this allows a high doping density to be used in the base, reducing the base resistance while maintaining gain. The main idea for creating SiGe HBTs is to include a graded germanium layer into the base of silicon BJT. As a result, higher current gain β, lower base transit time, which means higher f_t, and higher early voltage V_A, can be achieved.

3.4 SiGe HBT BICMOS

The objective of BiCMOS process integration is to create a final process with CMOS characteristics unchanged from the original CMOS only process and bipolar performance matching the bipolar only process. Combining both the high speed HBTs with the high integration capabilities of CMOS technology makes the technology suitable for high frequency integrated circuits applications and systems.

3.5 IHP's SiGe HBT BiCMOS Technology

3.5.1 Technology Introduction

As explained in [Cre08], IHP SG25 is a state-of-the-art 0.25 µm emitter strip width BiCMOS process, which benefits from the differing emitter-base structure to create HBT transitors. The implantation step for doping the extrinsic base layer or the elevated extrinsic base layer enhances the bipolar transistor's performance (f_t up to 180 GHz and f_{max} up to 200 GHz). IHP's SiGe BiCMOS technology is well suited for products and applications, such as, 60 GHz transceivers with ultra-high data rate.

3.5.2 Technology Passive Components

In addition to the HBT transistors and MOSFETS, the technology also supplies several passive components. In this section, the most important passive components are introduced.

3.5.2.1 Resistors

In integrated circuits, resistors are realized as thin film resistors, which consist of a thin metal or polysilicon layer between two isolation layers. The big advantage of this resistor is the small parasitic capacitances due to the large distance to the substrate. In the SG25 library, the devices are called rppd, rsil, and rhigh. They are implemented using the Gate-poly and Metal1 layers. Fig 3.2 shows the equivalent circuit model of a thin film resistor. L represents the parasitic connection wire inductance and R is the desired resistance. The components C_{Si}, C_{ox}, and R_{Si} are the substrate and substrate coupling parasitics.

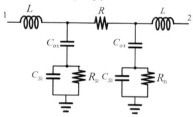

Fig 3.2 Resistor model

3.5.2.2 MIM Capacitors

Capacitors in integrated circuit design are frequently realized as MIM (Metal Insulator Metal) capacitors. In the SG25 library, they are named as *CMIM*. These capacitors have high quality factors and low tolerance. MIM capacitors can be modelled as shown in Fig 3.3. As can be noticed, the substrate coupling components influence only the lower plate of the capacitor. Therefore, only one side of the substrate parasitics is needed in the equivalent model.

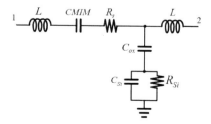

Fig 3.3 *CMIM* Model

3.5.2.3 Transmission Lines and Inductors

In the technology there exist several inductors with different values. Unfortunately, due to the high frequency of operation, the self-resonance frequency of the inductors limits their usage at such frequencies. Therefore, transmission lines had to be designed and simulated using electromagnetic simulation software, such as, ADS or sonnet. For the SG25 technology, [Hau09] presents the model of microstrip lines constructed between the second top metal (TM2) and metal1 layers. As a result of modeling, the R`L`G`C` parameters of the line were extracted and a scalable transmission line model was generated. As shown in Fig 3.4, the values of the inductance L` and resistance R` are frequency dependent mainly due to the skin effect, whereas for the capacitance C`= 77 pF/m and conductance G`= 0 S/m the values are frequency independent.

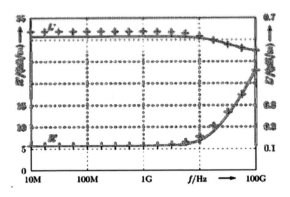

Fig 3.4 L` and R` for microstrip line of 300 μm length, (+) EM simulations, (-) tline model [Hau09]

For short transmission lines, the model in Fig 3.5 is used. The values of the model components are derived using the length of the line and the R`L`G`C` parameters by simple multiplication. The model in Fig 3.6 is used for inductors.

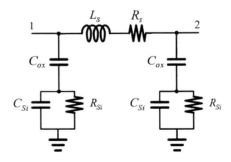

Fig 3.5 Inductive line model

Chapter 3. SiGe HBT BiCMOS Technology

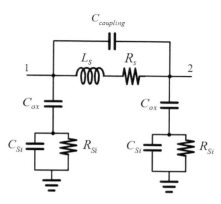

Fig 3.6 Inductor model

3.5.2.4 Varactors

Varactors are necessary components for voltage controlled oscillators and phase shifters. In the SG25 technology two types of varactors are available: Mvaricap, which is a conventional accumulation mode varactor, as will be explained in Chapter 5, and Svaricap, which is based on Schottky diodes. Because the quality factor of the varactor becomes worse at high frequencies, the choice of the varactor size and type becomes critical. Varactors' modeling is presented in Chapter 5.

Chapter 4 Charge-Pump-Based PLL System Modeling

4.1 Introduction

A voltage controlled oscillator (VCO) cannot be used directly to drive the modulator due to the frequency instability of its signal. In addition, a crystal oscillator can also not be used, because the crystals can only resonate at frequencies below 100 MHz. Therefore, a PLL is needed to stabilize the high frequency of the VCO.

In order to generate the 61.44 GHz carrier which is required by the transmitter, the well-known charge-pump-based (CP) PLL system was used. A PLL system is a closed loop system which compares the phases of two signals and generates a voltage proportional to the phase difference between them. This voltage is then filtered and applied to the control voltage line of a the VCO, which controls it to go faster or slower depending on the frequency and phase difference between the two signals.

A PLL system consists of several components:

- A VCO which generates a range of frequencies in dependence to the value of its control voltage input.
- A frequency divider which divides the frequency of the VCO into a signal with lower frequency which is suitable to be compared with the crystal oscillator signal.
- A phase detector (PD) or a phase frequency detector (PFD) that compares the phases of two signals and generates a signal that has an average voltage which is proportional to the phase difference between the two signals.
- A loop-filter which filters the higher harmonics of the output of the phase detector and generate a smoother voltage for driving the VCO voltage control line.
- A crystal oscillator (or the output of another PLL) that generates a high stability reference for the PLL.

This chapter begins by explaining some of the non-idealities in the output of a PLL. Then, the system architecture of the PLL is introduced. After that, the verilog-A model of the system is presented; this model helps in predicting the dynamic behavior of the system, such as, locking behavior and settling time. Then, several methods for predicting the phase noise of the system are explained. Finally, the trade-offs which a designer faces when designing a PLL are introduced. System simulations were used to optimize the filter parameters and resolve the trade-off between the loop dynamics, stability, spur power and phase noise performance, as shown in Chapter 9.

4.2 PLL Non-idealities

In an ideal case, the output of a PLL, which is based on LC VCOs, is a pure sinusoidal signal whose spectrum is a Dirac at the desired frequency, with no harmonics, spurs, or phase noise. In addition, in an ideal PLL, at the moment the PLL is switched on, the required frequency and signal should appear at the output in zero time. Unfortunately, this is not the case in practice, which necessitates discussing some of these non-idealities before going into the system architecture. Fig 4.1 shows the ideal spectrum of PLL at a frequency of $f_{osc.}$

Chapter 4. Charge-Pump-Based PLL System Modeling

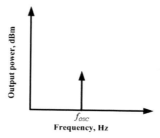

Fig 4.1 PLL ideal spectrum

4.2.1 Phase Noise

Phase noise is a measure of the deviation of the output signal from a perfectly periodic signal. Implemented circuits normally consist of noisy components, such as, transistors, resistors, and varactors. These components generate several kinds of noises, such as, shot noise, flicker noise, and thermal noise. Such noises affect the periodicity of the output signal of the oscillator. This effect is called phase noise and causes skirts to appear in the spectrum of the PLL instead of having a pure Dirac, as shown in Fig 4.2.

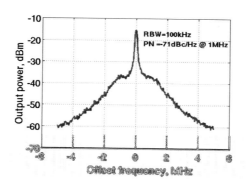

Fig 4.2 Phase noise non-ideality

4.2.2 Spur Power

In the output signal of a PLL power exists not only at the fundamental frequency of oscillation, but also at frequency offsets which are multiples of the frequency of the crystal oscillator from the carrier. These components in the spectrum are called spurs. Because the output of the phase detector contains many harmonics and the loop filter has a certain bandwidth, the loop filter cannot totally prevent these harmonics from affecting the control voltage line of the VCO. As a result, the control voltage of the VCO has a periodic voltage signal which causes frequency modulation at the output of the VCO. This effect is shown in Fig 4.3.

26 Chapter 4. Charge-Pump-Based PLL System Modeling

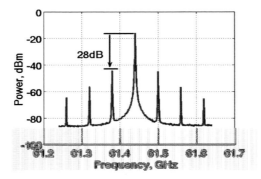

Fig 4.3 Spur non-ideality

4.2.3 Dynamic Behavior

Ideally, a PLL generates the desired output in zero time. Unfortunately, because a PLL is a closed loop control system, it takes some time for the loop to respond to a change in the frequency at the input or to settle to the desired frequency at the moment when the PLL is turned on. Therefore, it is important to understand the dynamic behavior of a PLL. The dynamic behavior of the PLL is controlled by the parameters of the components in the loop and can be analyzed by monitoring the tuning voltage of the VCO. It is noticed that, a change in the input frequency generates a transient on the control voltage which needs some time to finally settle to the voltage which corresponds to the new frequency. Such a transient is characterized by the settling time and overshoot. It also shows whether the output of the PLL is in the locked state or not, which indicates any stability problems in the behavior of the PLL. An example of a transient on the tuning voltage due to an input frequency step can be seen in Fig 4.4.

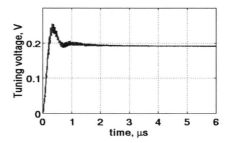

Fig 4.4 Tuning voltage transient

4.3 PLL Specifications

In the PLL specifications, two things had to be taken into account:

- The receiver uses a carrier recovery system which can follow the slow changes of the phase noise. This makes the bit error rate at the output of the receiver less sensitive to phase noise. Therefore, phase noise requirements are relaxed, [Rei05] and [Ulu10].

- The target system is a transceiver that works in a time division duplexing (TDD) mode with highly asymmetric traffic. As a result, the transmitter has to be turned off when the receiver is active (receiving control data, such as, commands to resend erratic frames for error correction) to minimize the transmitter feed-through into the receiver. In order to enable the transmission of high data rates up to 10 Gbps, in such conditions, the settling time of the PLL has to be fast. For clarification, assume that the device will receive control information after transmitting a certain number of frames or upon receiving a negative acknowledge NAK. Looking at Table 1.1, the frame length is around 76 µs. By making the settling time of the PLL faster, the duration of the PLL's settling time in comparison to the frame duration becomes smaller and the overhead is minimized. A 4 µs settling time was specified by the system designers to be sufficiently fast (5% of the frame length). As an example, receiving control information for NAK every 10 frames, which corresponds to a frame error rate of 10%, generates an overhead of 0.5% in the data rate (around 50 Mbps for a target of 10 Gbps).

By using these two considerations, the phase noise performance could be sacrificed and the loop bandwidth could be increased until the required settling time was achieved, taking into consideration the attenuation of the spurs.

4.4 System Architecture

The conventional CP-based PLL architecture, as shown in Fig 4.5, was adopted due to its lower risk and suitability for the required application. A common collector Colpitts voltage controlled oscillator (VCO) was used in a PLL system to generate the 61.44 GHz carrier. The VCO was followed by an output splitter which feeds the output of the VCO into two paths; one was used for the frequency divider and the other was used to drive the I/Q modulator. The frequency divider divides the frequency of the 61.44 GHz signal to a 60 MHz signal (the frequency of the crystal oscillator). The output of the divider was then fed into a conventional tri-state phase frequency detector (PFD), optimized to minimize the dead-zone. The output of the PFD was used to drive a highly matched charge pump (CP), which, with such matching, minimizes the power of the spurs generated around the carrier and reduces the noise generated by the charge pump transistors. Finally, the output of the charge pump was connected to a 2nd order integrated loop filter. The approach of capacitor multiplying was used to enable the integration of the loop filter on the same chip.

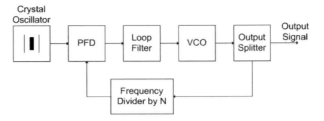

Fig 4.5 PLL block diagram

4.5 PLL Analysis Model

In order to enable the prediction of the PLL dynamic and noise behaviors, an analysis model can be generated by linearizing the components around an operating point [Bes03]. This operating point is the locked state. The variables in the linearized model are mainly excess phase and voltage quantities. Fig 4.6 shows the PLL's linearized model.

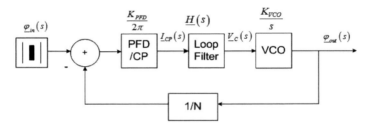

Fig 4.6 PLL linear model

The following explains the meaning of the different parameters and transfer functions in the model:

- The VCO gain (K_{VCO}): is the parameter that relates the change in the VCO output frequency in correspondence to a change in the VCO control voltage. Because the phase is the integration of the frequency, $1/s$ appears in the transfer function.
- The PFD gain (K_{PFD}): is the parameter that relates the change in the average output current at the output of the PFD/CP to the phase difference between the PFD input signals.
- $I_{CP}(s)$ is the charge pump current.
- The loop filter transfer function $\underline{H}(s)$, which is impedance that converts the CP current into the VCO control voltage. The conventional and integrated loop filter, which will be considered in this design are shown in Fig 4.7. The input impedance of the filter is given in (4.1).

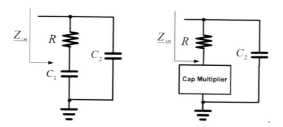

Fig 4.7 The conventional and integrated loop filters

$$\underline{Z}_{in} = \frac{RC_1 s + 1}{s^2 C_1 C_2 R + s(C_1 + C_2)} \qquad (4.1)$$

- The division ratio N: is the ratio between the output frequency of the VCO and the reference frequency.

From this model, a transfer function from the input to the output can be derived, which relates the output excess phase to input excess phase, as given in (4.3). This transfer function can be further investigated for the system stability, the loop bandwidth, and the dynamic behaviour.

$$\frac{\underline{\varphi}_{out}(s)}{\underline{\varphi}_{in}(s)} = \frac{\underline{G}_{forward}(s)}{1 + \underline{G}_{loop}(s)} \qquad (4.2)$$

$$\frac{\underline{\varphi}_{out}(s)}{\underline{\varphi}_{in}(s)} = \frac{K_{VCO} \dfrac{K_{PFD}}{2\pi}(1 + sC_1 R)}{s^2 N (C_1 + C_2)(1 + sC_s R) + K_{VCO} \dfrac{K_{PFD}}{2\pi}(1 + sC_1 R)} \qquad (4.3)$$

In the design of the PLL, the loop filter parameters are chosen in a way to allow for a PLL flat response, where the damping ratio is chosen to be unity, [Rog06].

4.6 Verilog-A Modeling

The dynamic behavior of a PLL can be simulated directly on circuit level. Unfortunately, this approach is time consuming. As a result, a system model using verilog-A was built to enable running the transient simulations efficiently. In this model, the code of each component was written separately and the hierarchy editor in cadence was used to enable the simulation of the whole system. See [Zou05], for more details about PLL verilog-A modeling. The verilog-A codes of the components can be found in appendix A.

4.7 Phase Noise Modeling

Because phase noise is an important parameter for wireless communications systems, as mentioned in Chapter 2, it is necessary to predict the phase noise performance of the PLL. By doing so, the optimization of the design in correspondence to the target application becomes possible. In this section, three different phase noise simulation methods are explained, as given in [Kun06]. Then, the frequency domain modeling method is elaborated.

4.7.1 Direct Simulation

The phase noise of a PLL system can be simulated directly on the circuit level using PSS and Pnoise analysis in SpectreRF. Unfortunately, this way of simulation consumes lots of time and resources and sometimes may end up with an error due to insufficient resources.

4.7.2 Time Domain Simulation

In this method, the phase noise of each of the components in the PLL is simulated using PSS and Pnoise analysis. Then, the result is converted into jitter which can be included in the verilog-A code and then the whole system is simulated and the jitter on the output is measured and converted back to phase noise.

4.7.3 Frequency Domain Simulation

This method uses the linearized model of the PLL, as shown in Fig 4.6. The noise generated by each component in the PLL is modeled as an additive noise at the output of the component. Then, under the assumption that the noise sources are uncorrelated, the superposition principle can be used to derive the transfer function of each noise source to the output. The model shown in Fig 4.7 can be used to predict the PLL's phase noise performance. The different transfer functions of the different noise sources can be derived, as shown in Table 4.1, and then the noise spectrum of each block can be simulated using cadence and shaped to the output using its own transfer function (Matlab can then be used to generate the predicted output spectrum). In this PLL architecture, the noise of the PFD/CP combination is the dominant part, as stated in [Rog06]. In this case, due to the integration of the loop filter, the loop filter also highly contributes to the PLL phase noise. It is also necessary to notice that by analyzing the block diagram in Fig 4.8 for each noise source, all noise sources are shaped to the output using a low pass filter characteristic, except the VCO noise, which is shaped using high pass filter characteristic. This makes the phase noise of the VCO at frequency offsets far from the carrier frequency more critical to the system than at smaller offsets.

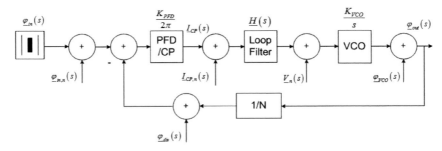

Fig 4.8 PLL noise model

Chapter 4. Charge-Pump-Based PLL System Modeling

Table 4.1 Noise and loop transfer functions [Kun06]

Forward gain	$\underline{G}_{forward}(s) = \dfrac{K_{PFD}}{2\pi} * \underline{H}(s) * \dfrac{2\pi K_{VCO}}{s}$
Reverse gain	$\underline{G}_{reverse}(s) = \dfrac{1}{N}$
Loop gain	$\underline{G}_{loop}(s) = \underline{G}_{forward}(s) * \underline{G}_{reverse}(s)$
VCO noise	$\dfrac{\underline{\varphi}_{out}(s)}{\underline{\varphi}_{vco}(s)} = \dfrac{N}{N + \underline{G}_{forward}(s)}$
Divider noise	$\dfrac{\underline{\varphi}_{out}(s)}{\underline{\varphi}_{div}(s)} = -\dfrac{\underline{G}_{forward}(s)}{1 + \underline{G}_{loop}(s)}$
Charge pump noise	$\dfrac{\underline{\varphi}_{out}(s)}{\underline{I}_{CP,n}(s)} = \dfrac{2\pi}{K_{PFD}} \dfrac{\underline{G}_{forward}(s)}{1 + \underline{G}_{loop}(s)}$
Loop filter noise	$\dfrac{\underline{\varphi}_{out}(s)}{\underline{V}_n(s)} = \dfrac{2\pi}{K_{PFD}} * \dfrac{1}{\underline{H}(s)} * \dfrac{\underline{G}_{forward}(s)}{1 + \underline{G}_{loop}(s)}$
Reference oscillator noise	$\dfrac{\underline{\varphi}_{out}(s)}{\underline{\varphi}_{in,n}(s)} = \dfrac{\underline{G}_{forward}(s)}{1 + \underline{G}_{loop}(s)}$

In order to clarify the high pass behavior of the VCO, a simplified model for the loop filter (series RC circuit) can be used and the division ratio can be set to unity. As a result, the VCO transfer function becomes:

$$\frac{\underline{\varphi}_{out}(s)}{\underline{\varphi}_{vco}(s)} = \frac{N}{N + \underline{G}_{forward}(s)} = \frac{1}{1 + \dfrac{K_{PFD}}{2\pi}\left(\dfrac{1}{sC_1} + R\right)\dfrac{K_{VCO}}{s}} \qquad (4.4)$$

Equation (4.4) can be written as:

$$\frac{\underline{\varphi}_{out}(s)}{\underline{\varphi}_{vco}(s)} = \frac{s^2}{s^2 + 2\zeta\omega_n s + \omega_n^2} \qquad (4.5)$$

where ζ is the damping ratio and ω_n is the natural frequency of a second order system.

Written in this form, (4.5) clearly shows a high pass behavior. But, because the frequency in the equation represents the frequency of phase changes around the locked frequency, it should be clear that, high pass, in this context, means frequencies far away from the locked frequency are passed to the output and not relative to zero.

4.8 Simulations

For the implementation of the PLL, the following parameters of the different blocks were used for system simulations: $K_{PFD} = I_{CP} = 230\ \mu A$ (for the integrated PLL) and $0.5\ mA$ (for the partly integrated PLL due to using larger capacitor), $K_{VCO} = 2 \times \pi \times 5$ Grad/V, $N = 1024$, and a 2nd order loop filter. The loop filter parameters were chosen to give the optimum loop bandwidth (optimally flat response) and settling time, as given in [Rog06]. For the targeted design, the loop bandwidth was chosen to be 1.6 MHz to enable the PLL to be as fast as required. The simulation results are shown in Chapter 9. The loop filter parameters which were used in the implementation of the PLL, for the three different bandwidths are as follows:

For a loop-bandwidth of 1.6 MHz, $C_1 = 110$ pF, $C_2 = 10$ pF, $R=10$ kΩ (integrated PLL).

For a loop-bandwidth of 1.0 MHz, $C_1 = 330$ pF, $C_2 = 33$ pF, $R=4$ kΩ (partly integrated PLL).

For a loop-bandwidth of 0.4 MHz, $C_1 = 2.2$ nF, $C_2 = 220$ pF, $R=2$ kΩ (partly integrated PLL).

4.9 PLL Trade-offs

The design of a PLL includes several trade-offs between, the dynamic behavior, the phase noise, spur power, and stability. Depending on the target application, some properties can be sacrificed to achieve better performance in others. The loop bandwidth, which is mainly controlled by the bandwidth of the loop filter, is a very important parameter in the design process, because it has to be chosen in a way to optimize the PLL according to the application. To explain the trade-offs, assume that a PLL with a fast dynamic behavior is required. To design the system for that purpose, the loop bandwidth has to be increased until the required speed is achieved. At the same time, by increasing the loop bandwidth, the noise of the charge pump and loop filter is shaped using a larger bandwidth to the output of the PLL, which degrades the phase noise behavior of the system. In addition, the harmonics on the tuning voltage line are less attenuated due to the larger loop bandwidth, which leads to an increased power in the spurs around the carrier. Matlab was used to analyze the effect of the loop bandwidth on the phase noise and settling time. For this purpose, the model presented in previous sections was used. Fig 4.9 shows the explained trade-off between phase noise and PLL dynamics. The bandwidths used here are only for clarifying the trade-off and are different from the values used in the design as in section 4.8.

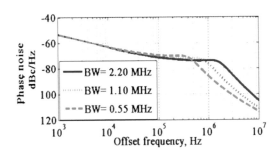

(a)

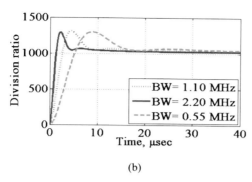

(b)

Fig 4.9 Loop bandwidth versus a) phase noise b) settling time

Matlab codes of that plots the dynamic behavior and the phase noise of the PLL can be found in Appendix C.

Chapter 5 Voltage Controlled Oscillators

5.1 Introduction

An oscillator is a circuit that generates a periodic waveform. Because oscillators generate waveforms without any input signal, they are typically able to amplify their own noise and build up oscillation. To achieve that, positive feedback must exist in the circuit. Voltage controlled oscillators (VCOs) are core components in wireless communication systems. They are oscillators, in which, the frequency of the output signal can be tuned using a control voltage input. For this purpose, a varactor is used. Varactors are variable capacitors, in which, the capacitance can be changed by applying different dc control voltages across them. Most commonly used VCOs in wireless communications are those which are built using LC-tanks and are called LC-VCOs. In this thesis, LC-VCOs will only be considered. In order to build an LC-VCO, a varactor is connected with an inductor (or transmission line which behaves inductively) to build the tank and the tank is then connected to an active circuit in order to compensate for the inherent losses in varactors and inductors and hence, sustaining the oscillation.

As was mentioned in Chapter 4, even though VCOs generate periodic signals, they cannot be used to directly drive the input of the modulator in a wireless communication system due to the poor frequency stability of the isolated VCO. Therefore, a VCO is normally placed in a PLL system to stabilize its frequency.

This chapter begins by introducing two different approaches that can be used in analyzing and designing an oscillator. After that, several components which can be used as varactors are presented and the properties of varactors, which should be considered by the designer before choosing a certain varactor for the VCO, are explained. Then, varactor modeling and the method to design test structures for varactors are presented. Following that, two kinds of LC-VCOs which are suitable for high frequency operation are introduced and the properties of VCOs that the designer should optimize according to the target application are explained. Finally, the working principle of quadrature VCOs is presented and the design of a 60 GHz common collector Colpitts QVCO is introduced.

5.2 Oscillator Models

In order to understand how to design oscillators, it is necessary to build models that can explain how an oscillation can build up in the circuit. This leads to conditions that should be satisfied in order to guarantee oscillation in the circuit at a specific frequency. An oscillator is a large signal nonlinear system, which means models that are based on small signal or linear models on circuit level are only beneficial in deriving conditions that guarantee the start up of oscillation in the circuit and giving an approximation for the oscillation frequency of the oscillator. For the large signal properties of an oscillator, different models and approaches have to be considered. In the following, two different linear oscillator models are introduced, namely, the positive feedback model and the negative resistance model.

5.2.1 The Positive Feedback Approach

In this model, the circuit is divided into two parts [Raz98], as shown in Fig 5.1:

- An amplifier with a transfer function $\underline{H}(j\omega)$.
- The tank (frequency selective network) with a transfer function $\underline{G}(j\omega)$.

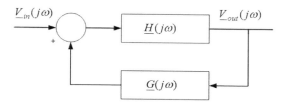

Fig 5.1 The positive feedback model

The transfer function of the system is:

$$\frac{\underline{V}_{out}(j\omega)}{\underline{V}_{in}(j\omega)} = \frac{\underline{H}(j\omega)}{1 - \underline{H}(j\omega) \cdot \underline{G}(j\omega)} \qquad (5.1)$$

If the input is assumed to be a noise impulse, the output of the system will grow indefinitely at a frequency ω_0 for which the loop gain is:

$$\underline{H}(j\omega_0) \cdot \underline{G}(j\omega_0) = 1 \qquad (5.2)$$

This gives the conditions that guarantee oscillation at a frequency ω_0 and are called the Barkhausen's criteria:

- Loop gain: $|\underline{H}(j\omega_0)| \cdot |\underline{G}(j\omega_0)| = 1$ (5.3)
- Loop phase: $phase(\underline{H}(j\omega_0) \cdot \underline{G}(j\omega_0)) = 2n\pi$ (5.4)

The phase condition determines the frequency of oscillation and the gain condition determines whether an oscillation can build up or not. These conditions are necessary but not sufficient for guaranteeing stable frequency of oscillation, as explained in [Kur69].

5.2.2 The Negative Resistance Approach

This approach also divides the circuit into two parts and finds the oscillation conditions by calculating the input impedance of each part [Ell07]. In an *RLC* circuit, if the circuit is excited with an impulse, the result is a damped oscillation (assuming *R* is positive) at the frequency where *L* resonates *C*. But, if the tank is included in a circuit that compensates its losses and generates a net negative resistance connected to the LC-tank, the result will be a growing oscillation, which is required for oscillators. Fig 5.2 can be used and the oscillation conditions are derived as:

$$IM(\underline{Z}_A) = IM(\underline{Z}_B) \qquad (5.4)$$

$$Re(\underline{Z}_A) = -Re(\underline{Z}_B) \qquad (5.5)$$

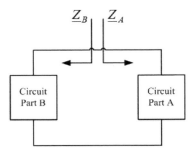

Fig 5.2 The negative resistance model

The imaginary parts determine the frequency of oscillation, whereas, the real parts determine the condition for an oscillation to build up.

5.3 Varactor Figure of Merits (FOMs)

As explained in [Gut06], in order to help the designer to choose a suitable varactor from the technology, several FOMs have to be defined, which allows the designer to compare the different available varactors and choose the most suitable one for the required application. The most important FOMs are the quality factor, the tuning ratio, and the self-resonance frequency.

5.3.1 The Quality Factor (Q-factor)

The Q-factor is a measure of the ability of a varactor to store energy efficiently. This is defined as the ratio between the stored and dissipated energy:

$$Q = 2\pi \frac{E_{stored}}{E_{loss}} \tag{5.6}$$

In terms of impedance, this can be transformed into:

$$Q = \frac{IM(Z_v)}{\text{Re}(Z_v)} \tag{5.7}$$

$$Q = \frac{1}{\omega C_v R_{loss}} \tag{5.8}$$

where Z_v is the input impedance of the varactor, R_{loss} is the varactor loss, and C_v is the varactor capacitance. Because the varactor is a part of the LC-tank in an LC-VCO, its Q-factor affects the bandwidth of the tank and hence, phase noise. In addition, the losses in the varactor can inject noise into the tank and degrade the phase noise performance. It is also important to notice that the Q-factor of a varactor becomes worse at higher frequencies, which makes it the dominant factor in determining the tank Q-factor at higher frequencies. This comes in contrast to lower frequencies, where the inductor's losses dominate the Q-factor.

5.3.2 The Tuning Ratio

The tuning ratio TR of a varactor affects directly the tuning range of a VCO and is defined as the ratio between the maximum capacitance C_{max} and the minimum capacitance C_{min} of the varactor:

$$TR = \frac{C_{max}}{C_{min}} \qquad (5.8)$$

5.3.3 The Self Resonance Frequency (SRF)

As the frequency of operation of VCOs goes higher, the effect of parasitic inductances in varactors becomes more critical. The SRF of a varactor is the frequency above which the impedance of a varactor becomes inductive and dominated by the series parasitic inductances. As a result, a varactor has to be used at frequencies well below its SRF. The SRF of a varactor is measured at C_{max}.

5.4 Varactor Types

As explained in [Gut06], several components can function as integrated varactors. These varactors can be divided into two categories: varactors which are based on MOSFETs, and others which are based on PN-diodes. Varactors can also be included in matrices, where multiple devices are connected to each other using a matrix structure. This gives the chance to increase the capacitance per unit area of the varactor.

5.4.1 PN-diode Varactors

PN-diode varactors are based on the idea of a PN-junction, where P-doped silicon is placed in contact with N-doped silicon, [Gut06]. The varactor structure can be seen in Fig 5.3. In order to operate the PN-junction as a varactor, the junction is reverse biased. This creates a depletion region capacitance, which is a function of the reverse voltage V_R across the junction terminals.

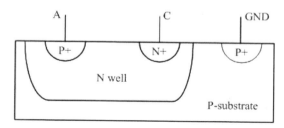

Fig 5.3 Diode varactor [Gut06]

As stated in [Ell07], a tuning ratio of 3-5 can be obtained using such varactors with quality factors higher than 30 at 5 GHz. Even though, PN-diode varactors are better than MOSFETs in terms of tuning ratio and quality factors as given in [Ell07], the available technology does not provide such varactors.

5.4.2 MOSFET Varactors

MOSFET varactors are based on metal-oxide-semiconductor structures. NMOS varactors are the most commonly used varactors in CMOS technologies, [Gut06]. The source and drain of the transistor are connected together S/D, and the capacitance between the gate G and S/D connection can be varied using the dc voltage applied across them. The capacitance of the varactor consists of the gate oxide capacitance per unit area C'_{ox} in series with the depletion capacitance per unit area C'_{si}. As a result, the total capacitance of a varactor can be calculated as:

$$C_{tot} = \left(\frac{1}{C'_{ox}} + \frac{1}{C'_{si}} \right) \cdot W \cdot L \qquad (5.9)$$

where W, and L are the width and length of the device, respectively. MOSFET varactors can be implemented in either accumulation or inversion mode.

5.4.2.1 Inversion Mode MOSFET

A standard MOSFET can function as a varactor, [Gut06]. Fig 5.4 shows a MOSFET transistor connected to function as a varactor. If a higher voltage than the threshold voltage is applied between the S/D and G, inversion occurs and the capacitance between the two terminals is a function of the voltage across them.

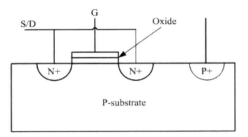

Fig 5.4 Inversion mode varactor [Gut06]

A drawback of this varactor is the existence of parasitic capacitances between the S/D and the substrate, which decreases the tuning ratio. Therefore, accumulation mode varactors are used more often.

5.4.2.2 Accumulation Mode MOSFET

The structure of accumulation mode varactors is shown in Fig 5.5, [Gut06]. The structure is similar to that of a standard MOSFET with the exception of placing the device inside an N-well, instead of the P-substrate. This reduces the S/D to substrate parasitic capacitances, hence, improving the tuning ratio in comparison to inversion mode varactors.

Chapter 5. Voltage Controlled Oscillators

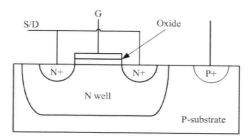

Fig 5.5 Accumulation mode varactor [Gut06]

The varactors used for the purpose of this thesis are based on this kind of varactors. As shown in section 5.5.1, the Q-factor of such varactors at 60 GHz can reach up to 4 with a tuning ratio around 3.

5.5 Varactor Modeling

Varactor modeling consists of two steps: first, is to build a circuit that explains the physical effects in the device and to derive equations that describe the parameters in this circuit. Second, is performing measurements for devices with different sizes to verify the model and to extract some parameters, which can only be found by measurements.

5.5.1 Simplified Circuit Model

For the purpose of this thesis, a simplified test structure of the varactor was designed and measured. Because the varactor is used in a differential VCO, a single ended characterization is sufficient due to the virtual grounds. As a result, the test structure in Fig 5.6 was used. A GSG probe and a network analyzer were used to measure the S-parameters. From the measured S-parameters, the parasitic pad capacitance and the parasitic transmission line connections were de-embedded and the varactor was modeled as a simple RC circuit, where the value of the capacitor and resistor at a specific frequency are voltage dependent. The measurement results are shown in Fig 5.7.

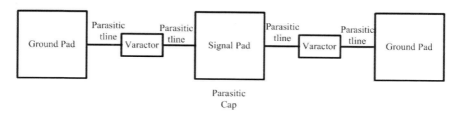

Fig 5.6 Varactor test structure

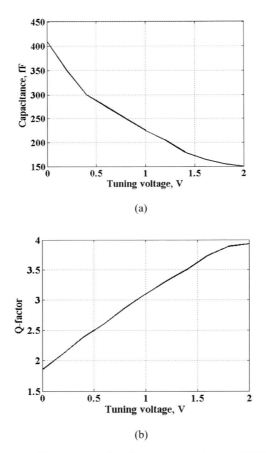

Fig 5.7 Capacitance and Q-factor as a function of tuning voltage at 60 GHz (a) Capacitance (b) Q-factor

For a more generalized modeling methodology of varactors, the following section can be followed.

5.5.2 Generalized Circuit Model

In order to enable the analysis of the FOMs of a specific varactor, a model should be generated that includes the parasitic effects in the device. By considering the physical effects in the device, a simplified linear model can be built using capacitors, resistors, and inductors. The focus of this section is on the accumulation mode varactors, because they are the most commonly used type. It is worth mentioning that, the principle of modeling is the same even for the other types of varactors. A model can be narrow-band and valid for only a small range of frequencies or broadband and valid for a larger range of frequencies. The model to be generated should have two important properties:

Chapter 5. Voltage Controlled Oscillators

- It should explain the physical effects behind the choice of components.
- It should be scalable, which means the validity of the model should be proven for devices with different sizes.

A simplified circuit model of the accumulation mode varactor is shown in Fig 5.8.

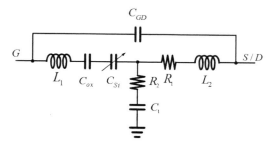

Fig 5.8 Accumulation mode varactor circuit model

where C_{ox} is the oxide capacitance, C_{GD} is the overlap capacitance between the gate and source/drain regions, C_{si} the depletion region capacitance, C_1 is the capacitance in the depletion region between the N-well and the substrate, R_1 is the N-well series resistance for the RF currents flowing into S/D terminal, R_2 is the N-well series resistance for the currents flowing into the substrate, L_1 is the inductance of the gate connection, L_2 is the inductance of the S/D connection. The analytical expressions of the model parameters are given in Table 5.1.

Table 5.1 Model parameters analytical expressions, [Gut06]

$C_{ox} = C_{ox}' A$	$L_1 = L_{01} + \dfrac{l_G l_1}{N}$
$C_{Si} = (c_a + c_i V_{var}) A$	$L_2 = L_{02} + \dfrac{l_{DS} l_2}{N}$
$C_{GD} = \dfrac{\varepsilon A_{overlap}}{d}$	$R_1 = W_p \dfrac{R_i - r_1 (V_{var} - V_{GS,off})}{NL_p}$
$C_1 = C_p + C_m A_1$	$R_2 = \dfrac{r_s}{A_1}$

where A is the area of the transistor, c_a is a parameter that the accumulation mode capacitance depends on, c_i is a parameter that the inversion mode capacitance depends on, V_{var} the voltage across the varactor terminals, $A_{overlap}$ is the overlap area between the S/D and the gate connections, d is the distance between the connection from the S/D region to the gate connection, C_p is a parasitic capacitance, C_m is the capacitance per unit area of the N-well and substrate depletion region, A_1 is the area between the substrate and N-well, L_{01} is the inductance of the gate connection to the varactor, L_{02} is the inductance of the S/D connection to the varactor, l_G is the inductance per unit of length of the metal connections on the gate of the varactor, l_{DS} is the inductance per unit of length of the metal connections on the S/D connection of the varactor, l_1 is the length of the connection to the gate of the varactor, l_2 is the length of the connection to the S/D terminal of the varactor, $V_{GS.off}$ is the voltage at which the N-well loses its free electrons, N is the number of transistors connected in parallel, r_s is the sheet resistance of the N-well, L_P is the gate length, W_P is the gate width, R_i is the inversion mode resistance of a unit varactor, r_1 is a parameter that controls the change of the resistance when it changes from inversion to accumulation.

5.5.3 Measurements and Testing

This step is necessary to approve that the model is working and to help in extracting some of the parameters, which can only be found by fitting measurements data to the model. As a result, this step should be done as accurately as possible. For this purpose, test structures are used. The role of the test structures is to interface the device to be tested to the measurement equipments while adding minimal parasitics. The most famous method for characterizing and modeling varactors is by using a vector network analyzer (NVA). In order to fully characterize the varactor, characterizing it as a two port system is preferred to one port characterization. The test system consists of the NVA, cables, connectors, bias-tees, and probes, as shown in Fig 5.9. After the test system is calibrated to eliminate the effect of cables, bias-tees, and probes, the S-parameters of the device under test (DUT) are measured. The short-open-load-through (SOLT) method is one way to perform this calibration. The device under test has to be connected to the measurement system. For that purpose, pads, guard ring, and metal layers are added. To eliminate the parasitic effects caused by them, the results should be de-embedded. To de-embedd these connections, their parasitic effects should be defined and eliminated by implementing several test structures. Fig 5.10 shows the different parasitic effects that are added to the DUT by pads and connections.

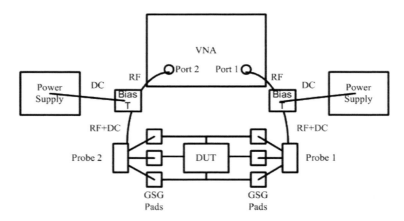

Fig 5.9 Measurement setup [Gut06]

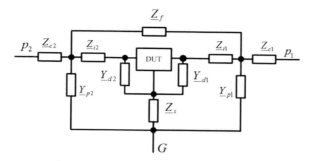

Fig 5.10 Setup parasitic [Gut06]

where Z_{c1}, and Z_{c2} are the ohmic losses of the contacts between the probes and test structure, Y_{p1} and Y_{p2} are the parasitic capacitances of the pads of both ports, Z_f is the coupling between the two ports, Z_{i1}, and Z_{i2} are the ohmic losses of the tracks connecting the pads to the DUT, Y_{d1}, and Y_{d2} are the parasitic capacitances between the tracks connecting the pads to the DUT and the ground, Z_s is the impedance between the DUT and the contact to ground on the substrate. As stated in [Agu04], for a well-designed test structure, Y_{d1}, Y_{d2}, Z_{i1}, Z_{i2} can be regarded as a part of the DUT, because the connections will be also needed in the real circuit. In addition, Z_f can be neglected. Furthermore, the effects of the substrate have an important influence when a varactor is inserted in a circuit. As a result, it should be taken as a part of the DUT, but if the structure is well grounded Z_s can be neglected. For full characterization, three different test structures are needed to de-embed the results and calculate the values of Z_{c1}, Z_{c2}, Y_{p1}, Y_{p2}, Z_{i1}, Z_{i2}, Y_{d1}, and Y_{d2}. The three structures are shown Fig 5.11. The single short is used to extract Z_{c1}, Z_{c2}. Then the single open is used to extract Y_{p1}, Y_{p2}. After that, a short is used to extract Z_{i1}, Z_{i2}. Finally, an open structure is used to extract Y_{d1}, Y_{d2}. For more details about the mathematics of the de-embedding process, see [Gut06].

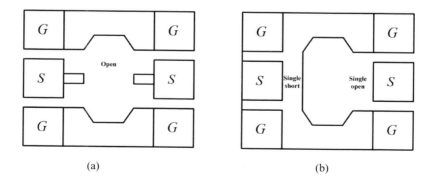

(a)   (b)

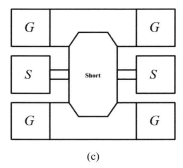

(c)

Fig 5.11 (a) Open (b) Single open, single short (c) Short, [Agu04]

5.6 VCO Properties

In this section, three important VCO properties are considered: the tuning range, the phase noise, and the output power.

5.6.1 Tuning Range

The tuning behavior of a VCO relates the change in the output frequency of the VCO to the change in its tuning voltage control input. It depends on the tuning behavior of the varactor and the VCO architecture (the effect of the architecture parasitics on the varactor). Equation (5.10) shows the frequency change in response to the capacitance of varactor.

$$\omega = \frac{1}{\sqrt{LC}} \qquad (5.10)$$

where L is the tank inductance, and C is the total capacitance of the varactor and the parasitic.

The tuning behavior is important for system considerations, specifically the linearity of the tuning curve and the value of the VCO gain, because they affect the stability and the dynamic behavior of the PLL.

5.6.2 Phase Noise

As shown in [Raz03], the phase noise of the VCO is shaped by a high pass filter to the output of the PLL, which makes the VCO noise at frequency offsets far from the carrier more critical to the system than the noise at low frequency offsets. In order to optimize the VCO for better phase noise performance, a phase noise model should be developed to help the designers in controlling the circuit parameters. To serve this purpose, two different models were developed, namely: Leeson's model and the linear time variant (LTV) model. In the following, the two models are introduced, as presented in [Haj98].

5.6.2.1 Leeson's Model

In this model [Haj98], the circuit is treated as a linear time invariant (LTI) system and the noise of the tank losses is shaped to the output by the tank impedance. As a result of this analysis, the phase noise is calculated as:

Chapter 5. Voltage Controlled Oscillators

$$L(\Delta\omega) = 10\log\left[\frac{2kT}{P_{sig}}\left(\frac{\omega_0}{2Q\Delta\omega}\right)^2\right] \quad (5.11)$$

where $L(\Delta\omega)$ is the phase noise at offset $\Delta\omega$ from the carrier, k is Bolzmann's constant, T is the absolute temperature, ω_0 is the frequency of oscillation, P_{sig} is the power dissipated in the tank losses resistor, and Q is the loaded quality factor of the tank.

Unfortunately, this equation does not accurately describe the measured spectrum in reality. In the measured spectrum, two additional regions other than the $1/(\Delta\omega)^2$ region are observed: the $1/(\Delta\omega)^3$ region and the noise floor. As a result, (5.11) has to be modified to (5.12), which includes the other two effects.

$$L(\Delta\omega) = 10\log\left[\frac{2FkT}{P_{sig}}\left\{\left(1+\frac{\omega_0}{2Q\Delta\omega}\right)^2\right\}\left(1+\frac{\Delta\omega_{1/f^3}}{\Delta\omega}\right)\right] \quad (5.12)$$

where F is a fitting factor which depends on the device noise figure, and $\Delta\omega_{1/f^3}$ is the $1/f$ noise corner frequency.

Equation (5.12) shows that increasing the output power and the quality factor of the tank improves the phase noise performance. Unfortunately, the factor F in the equation is an empirical factor and unknown to the designer at the moment of the design, which may be misleading sometimes e.g. using an active inductor to boost the inductor's quality factor. Increasing Q is beneficial, but using an active circuit for that purpose increases F at the same time, which might lead to worse phase noise even though Q is improved. In addition, the $\Delta\omega_{1/f^3}$ corner in measured spectrums does not really correspond to the $1/f$ noise corner frequency. Moreover, the model cannot really explain the up-conversion of the $1/f$ noise to frequencies around the carrier. As a result, it is concluded that, this model can provide qualitative description for phase noise, but is unable to quantitatively predict phase noise. Therefore, a model that can predict the phase noise of the circuit quantitatively and give better insight into controlling the design parameters for phase noise optimization is required. The LTV model solves the shortcomings of Leeson's model and gives better insight into how noise is transformed into phase noise.

5.6.2.2 LTV Model

This model was presented in [Haj98]. In this model, the circuit is treated as an LTV system. Before going into the details of this model, the linear and time variant behavior of an oscillator should be explained. This can be done by considering the behavior of the phase of the output signal when a noise impulse is injected into the circuit, as shown in Fig 5.12 for an LC-tank.

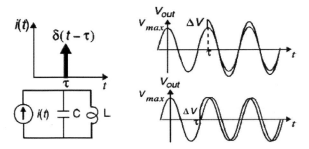

Fig 5.12 The effect of a noise impulse on the output waveform of an oscillator [Haj98]

- Time variant: a system is regarded as time variant, when the response of the system to a signal depends on the time at which the signal is applied to the system. By applying a noise impulse into an oscillator and studying the effect on the resulting phase shift in the output signal, the time variant behavior of the system can be proven. Because this phase shift is dependent on the moment at which this impulse is injected in the period of oscillation, the system is regarded as time variant. For example, the phase shift is maximum if the impulse occurs at the zero crossings of the waveform, whereas, it is minimum at the peaks of the waveform.
- Linearity: a system is regarded as linear if it satisfies the superposition principle, as given in (5.13):

$$\begin{matrix} X_1(t) \rightarrow Y_1(t) \\ X_2(t) \rightarrow Y_2(t) \end{matrix} \Rightarrow aX_1(t) + bX_2(t) \rightarrow aY_1(t) + bY_2(t) \qquad (5.13)$$

System linearity depends on the input and output variables under consideration. In this case, even though oscillators are non-linear systems when voltage is considered, the phase shift resulting from a noise impulse is linear. This can be proven by applying impulses with different areas (charges) and noticing the phase shift as a function of this charge.

Because an oscillator is an LTV system, it can be fully described by its impulse response $h_\phi(t,\tau)$, which is related to the sensitivity of the output signal to impulses at different time moments in the period of oscillation and can be found by simulating the circuit for each noise current in the circuit. The excess phase $\varphi(t)$ of the output signal in presence of noise source $i(t)$ can be calculated by using the superposition integral as:

$$\phi(t) = \int_{-\infty}^{\infty} i(t) h_\phi(t,\tau) dt \qquad (5.14)$$

In order to calculate the phase noise from this excess phase, Fourier series can be used to convert the impulse response to Fourier series coefficients and then substitute the value of the phase into the phase of a sinusoidal to calculate the phase modulation. This was done in [Haj98], and the result of the calculations is:

Chapter 5. Voltage Controlled Oscillators

$$L\{\Delta\omega\} = 10\log\left(\frac{\sum_k C_k \left[\frac{i_n^2(k,\Delta\omega)}{\Delta f}\right]}{4\Delta\omega^2}\right) \quad (5.15)$$

where C_k are the Fourier coefficients of the impulse response, and $\overline{\frac{i_n^2(k,\Delta\omega)}{\Delta f}}$ is the noise source current density at the kth harmonic of the oscillation frequency.

In addition, the results of the calculation give an estimation of the corner frequency in the spectrum as given in (5.16)

$$\omega_{1/f^3} = \omega_{1/f}\left(\frac{C_0}{C_1}\right)^2 \quad (5.16)$$

where $\omega_{1/f}$ is the flicker noise corner frequency, C_0 is the dc Fourier coefficient of the impulse response, and C_1 is the Fourier coefficient of the fundamental mode of the impulse response.

In addition to analyzing the effect of stationary noise sources on phase noise, the LTV theory gives also the ability to analyze the effect of cyclo-stationary noise sources, such as, the shot noise of the collector current. This can be done by defining a new impulse response of the circuit from the impulse response $h_\phi(t,\tau)$ which takes into account the periodicity of the noise source.

$$h_{\phi,cyclo}(t,\tau) = h_\phi(t,\tau)\alpha(t) \quad (5.17)$$

where $\alpha(t)$ is a periodic function with a maximum of 1, and can be derived from the currents or voltages waveforms of the oscillator. By using the coefficients of $h_{\phi,cyclo}(t,\tau)$ instead of $h_\phi(t,\tau)$ in equation (5.14) the phase noise of the cyclo-stationary noise source can be calculated. More details can be found in [Haj98] and [Mar99].

5.6.3 Output Power

The output power of the VCO is an important parameter, because it needs to drive an output splitter, which, in turn, drives the modulator. The modulator requires a certain power level to function properly. In [Rog03], several theoretical approximations of the output power of an oscillator are given. In this thesis, such approximations were used only as a first step for the design. Then the simulator was used for further optimization.

5.7 Types of LC VCOs

As was previously mentioned, LC-VCOs consist of a tank and an amplifier. In integrated circuits, the most common types of LC-VCOs are the cross coupled and the Colpitts VCOs.

5.7.1 Cross Coupled VCO

The cross coupled VCO is shown in Fig 5.13, [Bar12A]. In order to understand its behavior at high frequencies, the negative conductance approach is used for the circuit analysis. Fig 5.14 shows a small signal model for the VCO including the high frequency parasitic, such as, the base resistance R_b, base-emitter capacitance C_{be}, the base-collector capacitance C_{bc}, and the emitter resistance R_e. First, the circuit is analyzed by considering a short circuit in-place of C_m, and then, the effect of C_m is taken into account.

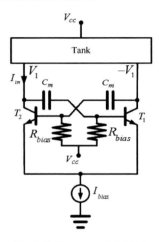

Fig 5.13 Cross coupled VCO

The negative conductance of the oscillator can be defined as:

$$G_{neg} = \text{Re}[\underline{Y}_{in}] = \text{Re}[\frac{1}{\underline{Z}_{in}}] = \text{Re}\left[\frac{I_{in}}{2V_1}\right] \quad (5.18)$$

In literature [Rog03], [Jeo06], equation (5.19) is commonly used in estimating the negative conductance G_{neg} of the oscillator:

$$G_{neg} = -\frac{g_m}{2} \quad (5.19)$$

where g_m is the transistor trans-conductance.

Unfortunately, this is not valid for high frequencies, due to the effect of transistor parasitics, such as, the base resistance R_b, the base emitter capacitance C_{be}, the base collector capacitance C_{bc}, the collector resistance R_c (not shown in Fig 5.14), and the emitter resistance R_e. Even at low frequencies and small transistor size, the estimation is inaccurate due to R_e. In the following, the effect of the parasitics and C_m is analyzed:

Chapter 5. Voltage Controlled Oscillators

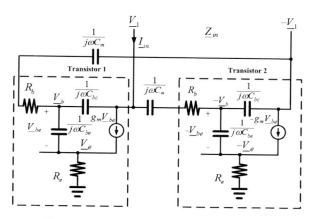

Fig 5.14 Cross coupled VCO small signal model

1. The effect of R_e: by neglecting all other parasitics and replacing C_m with a short circuit, (5.20) can be derived. It can be concluded that R_e reduces G_{neg} by a factor of $1/(1+g_m R_e)$ for low frequencies.

$$G_{neg} = -\frac{g_m}{2}\left(\frac{1}{1+g_m R_e}\right) \qquad (5.20)$$

2. The effect of C_{be} and R_b: (5.19) predicts a flat negative conductance over the whole frequency range, in reality the magnitude of the negative conductance decreases with frequency. This is explained by analyzing the circuit taking into account C_{be} and R_b, while neglecting the other effects. As a result, (5.19) has to be modified to:

$$G_{neg} = \frac{1}{2}\left(\frac{-g_m + \omega^2 R_b C_{be}^2}{\omega^2 C_{be}^2 R_b^2 + 1}\right) \qquad (5.21)$$

As can be noticed, a pole, which decreases the magnitude of the negative conductance, appears in the equation. The pole-frequency is:

$$f_{pole} = \frac{1}{2\pi R_b C_{be}} \qquad (5.22)$$

the G_{neg} crosses zero at a frequency of:

$$f_{zero,neg} = \frac{1}{2\pi}\sqrt{\frac{g_m}{C_{be}^2 R_b}} = \sqrt{f_t f_{pole}} \qquad (5.23)$$

where f_t is the transit frequency of the technology.

3. Taking the effect of C_{be}, R_b, and C_{bc} into account, the expressions for G_{neg} and f_{pole} are further modified as:

$$G_{neg} = \frac{1}{2}\left(\frac{-g_m + \left((C_{tot}+C_{bc})^2 + C_{bc}C_{tot}g_m R_b\right)R_b\omega^2}{1+\omega^2 R_b^2 C_{tot}^2}\right) \quad (5.24)$$

$$f_{pole} = \frac{1}{2\pi R_b C_{tot}} \quad (5.25)$$

where $C_{tot} = C_{be}+C_{bc}$. From (5.24), an estimation of the upper limit of the negative conductance can be derived as a function of the technology parameters. The designer can use such an upper limit to have a first judgment about the feasibility of using the cross-coupled oscillator for the target frequency, by comparing the tank losses with the upper limit. The upper limit of G_{neg} is:

$$G_{neg} = -\frac{1}{2}\left(\frac{g_m}{1+\omega^2 R_b^2 C_{tot}^2}\right) \quad (5.26)$$

and the G_{neg} crosses zero at a frequency of:

$$f_{zero,neg} = \frac{1}{2\pi}\sqrt{\frac{g_m}{\left((C_{tot}+C_{bc})^2 + C_{bc}C_{tot}g_m R_b\right)R_b}} \quad (5.27)$$

4. The effect of feedback capacitance on the G_{neg} of the oscillator: in order to boost the ability of cross-coupled oscillator to give negative conductance at higher frequencies and to decouple the bias from the outputs, a capacitor C_m is commonly inserted in the feedback path to eliminate the effect of C_{bc}, as shown in Fig 5.13. The negative conductance for this case can be derived as:

$$G_{neg} = \frac{\left(\frac{-g_m(C_m-C_{bc})}{(C_{tot}+C_m)}\right)+\left(\frac{r_b C_{tot} C_m^2 (C_{tot}+C_{bc}+C_{bc}r_b g_m)+(C_{tot}+C_m)r_b C_m^2 C_{bc}}{(C_{tot}+C_m)^2}\omega^2\right)}{1+\left(\frac{C_{tot}C_m}{C_{tot}+C_m}\right)^2 r_b^2 \omega^2} \quad (5.28)$$

Circuit analysis shows that the pole-frequency shifts to a frequency of:

$$f_{pole} = \frac{1}{2\pi R_b \left(C_{tot}C_m/(C_{tot}+C_m)\right)} \quad (5.29)$$

The upper limit of negative conductance is:

$$G_{neg} = -\frac{1}{2}\frac{g_m(C_m - C_{bc})/(C_{tot} + C_m)}{1 + \omega^2 R_b^2 \left(\frac{C_m C_{tot}}{C_m + C_{tot}}\right)^2} \quad , C_m > C_{bc} \quad (5.30)$$

and the G_{neg} crosses zero at a frequency of:

$$f_{zero,neg} = \frac{1}{2\pi}\sqrt{\frac{g_m(C_m - C_{bc})(C_m + C_{tot})}{(C_{tot} + C_{bc})C_m^2 R_b \left(C_{tot} + C_{bc} + g_m R_b \left(C_{tot} C_{bc}/(C_{tot} + C_{bc})\right)\right)}} \quad (5.31)$$

The previous analysis bring a lot of insight into the limitations of the cross coupled VCO at high frequencies.

1. Choosing C_m lower than C_{bc} eliminates the ability of the circuit to give negative conductance, as can be seen in (5.30).
2. A maximum limit of the ability of the circuit to generate negative conductance is given in (5.26).
3. A maximum frequency at which the circuit can generate negative conductance is given in (5.27) and (5.31).
4. Even though the maximum frequency previously derived does not prove oscillation feasibility, because the negative conductance could be of small magnitude, it does prove oscillation inability.

The analytical derivations were validated and illustrated by simulations. The simulation results correspond to a transistor with the following parameters: $R_b = 17\ \Omega$, $C_{bc} = 18$ fF, $C_{be} = 195$ fF, $R_e = 3.2\ \Omega$, $R_c = 11.4\ \Omega$, and $g_m = 257$ mS. However, the analysis results were successfully validated for several other transistor sizes, all available in the same IC technology. As can be noticed in Fig 5.15, $f_{zero,neg}$ taking all effects into account is around 65 GHz. When only C_{be} and R_b are taken into account (equation (5.23)) the estimated $f_{zero,neg}$ is 100 GHz, when C_{bc} effect is added (equation (5.27)), the estimated $f_{zero,neg}$ is 76 GHz. The effect of C_m was also evaluated. As can be noticed from Fig 5.16, where the conductance of the circuit is plotted versus frequency for different values of C_m, using $C_m = 17$ fF $\cong C_{bc}$ generates positive values for the conductance, as closely predicted by (5.30). Using larger values for C_m will finally bring the conductance to the short circuited case, as in (5.24). It can also be deduced that, even though using C_m improves the magnitude of negative conductance at higher frequency, the magnitude can still be low and insufficient for compensating the tank losses. Simulations show that a feedback capacitance in the range of $2C_{bc}$ to $3C_{bc}$ is a good beginning for the design process. The value can then be further optimized using simulations. Unfortunately, (5.28) gives results which is within 30% around the accurate value. This is mainly due to the high effect of the collector resistance and the substrate capacitance at high frequencies. As a result, simulations were only considered, in the case of using C_m in the circuit, to confirm the effect of using C_m smaller than C_{bc} on the negative conductance and to show what value of C_m the designer can choose at the beginning of the design process.

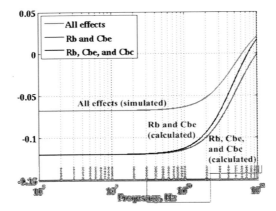

Fig 5.15 Simulated (all effects curve) and calculated parasitic effects (the other two curves) on the negative conductance of the cross-coupled oscillator using (5.21) and (5.24)

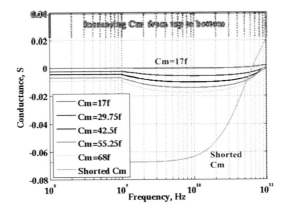

Fig 5.16 Simulated effect of C_m on the negative conductance of the cross coupled oscillator

5.7.2 Common Collector Differential Colpitts VCO

Colpitts VCOs are well-known for their excellent phase noise performance, especially at higher frequencies, in comparison to the cross coupled topology. This is mainly due to the Q-factor improvement of the varactor using the capacitive divider, which generates a higher Q-factor of the tank, especially that, at higher frequencies the varactor Q-factor is the dominant loss in the LC-tank. As a result, the common collector Colpitts differential VCO shown in Fig 5.17 was adopted, [Bar10A].

5.7.2.1 Design Trade-offs

Before beginning the design process, a designer should be familiar with the different design challenges and trade-offs in the design of Colpitts VCOs. Understanding these trade-offs helps the designer to optimize the VCO according to the target application. In this section, these trade-offs are explained.

- Differential versus single-ended: differential architectures are superior to single-ended ones when a component is to be used inside a system. Because a VCO cannot be used as a standalone component without a PLL due to its high jitter and phase noise, it is necessary to choose a differential architecture, even though a single-ended consumes lower power. This choice minimizes the substrate noise and power supply variation effects by rejecting common mode noise. In addition, it makes the physical design process much simpler due to easier grounding requirements, because of the virtual grounds in differential circuits. In designing a differential Colpitts VCO, it is very important to avoid common mode oscillation by choosing the correct range of the transistors' transconductance, as shown in [Yod07] for the common base differential Colpitts VCO.

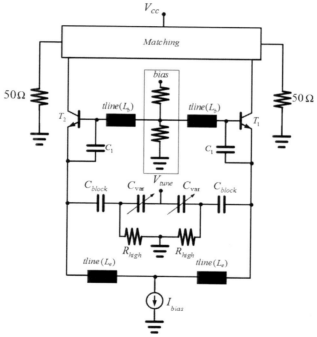

Fig 5.17 Differential common collector Colpitts VCO [Bar10A]

- Tuning range versus phase noise performance: the oscillation frequency of the VCO, as given in [Li03a], can be approximated by

$$\omega_{osc} = \frac{1}{\sqrt{LC_{eff}}} \qquad (5.32)$$

where $C_{eff} = C_{bc} + \dfrac{(C_{be}+C_1)C_{equ}}{C_{equ}+C_{be}+C_1}$ and C_{equ} is the quivalent capacitance on the emitter node. As can be seen from (5.26), the oscillation frequency is highly affected by the parasitic capacitances of the transistor, mainly, the base-emitter and the base-collector capacitances, which reduces the tuning range. As a result, using a larger transistor reduces the tuning range. But in cases where low phase noise is more important than the tuning range, a larger transistor size is beneficial, because the phase noise of the VCO is very sensitive to the transistor's base resistance and the base resistance is reduced by using a larger transistor. Another important issue is the varactor size which affects both the tuning range and the phase noise and should be chosen carefully according to the application.

- Output power and phase noise versus power consumption: for the phase noise of the VCO, there exists an optimum bias current density. In addition, the bias current of the circuit controls the negative resistance and the output power of the VCO. As a result, the bias current should be carefully chosen to generate the required negative resistance and output power with the best possible phase noise performance and minimum power consumption.

5.7.2.2 The effect of Parasitics and Load Resistance on the Negative Resistance

A half circuit small signal model of the VCO is shown in Fig 5.18. It should be emphasized that this model is used only to explain the effect of the parasitics on the VCO negative resistance and not for exact values of the negative resistance. For exact values, the circuit in Fig 5.17 should be investigated.

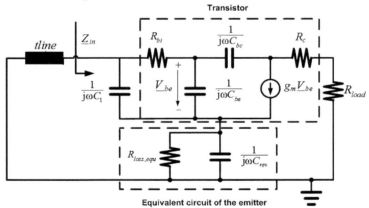

Fig 5.18 A half circuit small signal model of the VCO

In Fig 5.18, the capacitor C_{equ} and the resistor $R_{loss,equ}$ model the impedance on the emitter of the VCO in Fig 5.17 at a specific frequency, as shown in Fig 5.19. For the used transistor and other components, the following values apply:

Chapter 5. Voltage Controlled Oscillators

R_{bi} = 15 Ω, C_{be} = 146 fF, C_{bc} = 20 fF, R_C = 10 Ω, g_m = 170 mS, $R_{loss,var}$ = 133 Ω, C_{var} = 95 fF, $R_{loss,equ}$ = 182 Ω, C_{equ} = 28 fF, C_1 = 65 fF, C_{block} = 300 fF, $R \gg R_{loss,var}$, L_e = 159 pH, L_b = 105 pH.

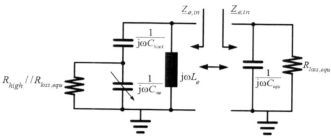

Fig 5.19 An equivalent half circuit of the emitter node at a specific frequency

In literature [Rog03], the negative resistance of Colpitts VCOs is frequently approximated by:

$$Re\{\underline{Z}_{in}\} = R_{neg} = -\frac{g_m}{C_{total} C_{equ} \omega^2} \qquad (5.33)$$

where $C_{total} = C_1 + C_{be}$. This approximation is good for low frequency designs. However, for higher frequencies it tends to yield imprecise results, e.g. since the varactor losses represented by $R_{loss,var}$ and the transistor feedback capacitance C_{bc} are not included. Hence, we present a new expression for the negative resistance including $R_{loss,equ}$ and C_{bc}:

$$R_{neg} = \frac{-g_m C_{equ} C_{total} + \dfrac{C_{total}^2}{R_{loss,equ}}}{((C_{total} + C_{equ})C_{bc} + C_{equ} C_{total})^2 \omega^2 + (g_m C_{bc} + \dfrac{C_{total}}{R_{loss,equ}})^2} \qquad (5.34)$$

Of course, the expression simplifies to (5.33) if C_{bc} = 0 and $R_{loss,equ}$ = ∞. Furthermore, a considerable impact of the transistor base resistance R_{bi}, and the transistor collector resistance R_C connected in series with the VCO load resistance R_{load} was observed. It would be possible to include also R_{bi} and R_{load} into (5.34). However, in this case, the corresponding equation becomes very complicated and gives not much transparent insights. Hence, CAD simulations to evaluate the impact of all elements, was performed. The results are plotted in Fig 5.20. Using the previously given parameters, the calculation of negative resistance using (5.33) at 60 GHz gives -203 Ω. Taking $R_{loss,var}$ into account gives -121 Ω. Equation (5.34) gives -20 Ω. Finally, taking all effects including R_{load} = 33 Ω and R_C = 10 Ω into account results in -5 Ω.

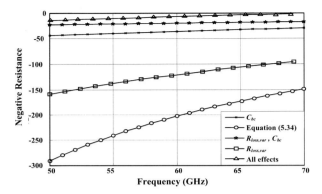

Fig 5.20 Calculated and simulated effect of parasitics on the negative resistance of the VCO

It can be concluded that the commonly used (5.33) is only suitable as a first order estimation to give the designer a feeling regarding the element values in a specific frequency range. The estimations can be enhanced by using (5.34) and should finally be supported by CAD simulations considering the original circuit in Fig 5.17.

5.7.2.3 Components Sizing

In this section, the design process and the sizing of the different components in the VCO are presented.

- **Varactor**

At 60 GHz, the quality factor of the varactor becomes the dominant loss of the tank. As a result, an accurate varactor model at 60 GHz is necessary. It was shown in [Rog03], that the higher C_{total}/C_{equ} the better the VCO phase noise. However, the larger the ratio the lower the negative resistance. Therefore, the ratio has to be limited. As a result, a first step in the design process is to study the available varactors, their capacitance range and quality factor and then use equation (5.33) or (5.34) to estimate the required capacitance. The varactors in this technology are accumulation mode varactors with C_{max}/C_{min} around 3 and a quality factor around 3 at 60 GHz.

- **Transmission Lines**

Typically, at 60 GHz, the use of spiral inductors is limited by their self-resonance frequency. Hence, a grounded transmission line is used. If the length l is shorter than a quarter of the wave length λ, the equivalent value of the inductor can be calculated by [Che83]:

$$L_{equ} = \frac{Z_o}{\omega} tan(2\pi \frac{l}{\lambda}) \quad (5.35)$$

where Z_o is the characteristic impedance of the transmission line. The transmission lines were simulated in ADS-Momentum. Due to the difficulty in using cadence PSS analysis with distributed elements models, the ADS results were fitted to new models using the analogue environment transmission line modeller which can generate lumped elements models that can be used with PSS and Pnoise analysis in cadence. More details about transmission lines' modelling can be found in [Hau09]. A second step in the design is to study the losses of the transmission lines at the frequency of interest. This gives the designer a feeling about the range of values of the required negative resistance. A simulation of the losses of transmission lines for inductances in the range of 85 to 140 pH shows losses $R_{ind,loss}$ which range between 2 to 4 Ω at 60 GHz (2.5 Ω for the used transmission line 105 pH).

The transmission lines are used with different l at several nodes:

1) For the transmission line in the emitter path, it should be noticed that the emitter should see capacitive impedance for the circuit to generate negative resistance. As a result, l should be chosen to resonate with the series combination of the varactor and C_{block} on a much lower frequency than the frequency of interest. Without using this transmission line, two current sources will have to be used, one for each half, eliminating the chance to get rid of the current source parasitic capacitance by making its node a virtual ground.

2) The base transmission line controls the oscillation frequency of the oscillator and is chosen to eliminate the imaginary capacitive part of the input impedance at the oscillation frequency. The losses of this line determine the magnitude of the required negative resistance.

3) The transmission lines in the matching network are chosen to transform the 50 Ω load to the optimum load impedance.

- **Transistor**

Because the transistor size affects phase noise, tuning range and output power, choosing a certain size is an iterative process. The process depends on the optimization goal of the VCO, as explained in the design trade-offs in section 5.7.2.1.

- **Load Impedance**

One of the important reasons of choosing a common collector Colpitts VCO is its ability to isolate the tank from the load which reduces the tank losses and improves the performance and results in lower power consumption due to eliminating the output buffer. At high frequencies, the effect of the load on the negative resistance is critical. As a result, the question of the optimum load resistance comes into role, specially that the output resistance of the VCO may be negative which makes small signal conjugate matching not possible, moreover, a large signal matching can lead to a resistance value which eliminates the negative resistance in the input impedance. Hence, the choice of the optimum load is not straight forward.

Fig 5.21 shows the effect of changing the load resistor on the negative resistance of the VCO. Assuming a linear decrease in the large signal transconductance of the VCO, makes (5.36) optimum for the choice of maximum power delivered to the tank, as given in [Ell07].

$$R_{neg} = -3R_{ind,loss}. \quad (5.36)$$

where $R_{ind,loss}$ is the base transmission line loss. As a result, the load resistor can be swept to find the optimum for the whole tuning range. Then, transient simulations are used for further optimization. Taking into consideration the base transmission line losses, an optimum load resistance of 33 Ω in parallel with a 210 pH inductance for compensating the capacitive output of the transistor was found to be optimum. A matching network, which takes into account a 45 fF for the pad capacitance, was designed to transform the 50 Ω resistance into the optimum impedance over the tuning bandwidth.

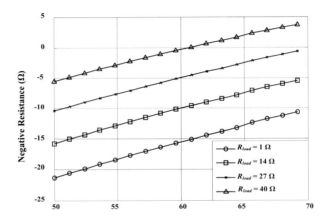

Fig 5.21 Simulated negative resistance for different load values

5.8 Common Collector Colpitts Quadrature VCO (QVCO)

In order to maximize the data rates in wireless communication systems, it is necessary to employ higher order modulations and this, in turn, requires the generation of high quality quadrature signals. In addition, quadrature signals are used in image rejection transceivers' architectures. In open literature, several different approaches to the generatoin of quadrature signals have been reported. [Her03] discusses the generation of quadrature signals using a VCO with a static frequency divider; in this case, using static dividers with a 120 GHz VCO is a very difficult approach to follow at such a high frequency with the available technology f_t. In [Ell09], a QVCO based on the cross coupled VCO as the core oscillator was introduced; this QVCO does not provide a suitable phase noise performance for some high-end applications. In [Yod06], a QVCO based on a common base colpitts VCO was proposed. In [Hac03], a 28 GHz QVCO is introduced. In this thesis, a QVCO based on the common collector Colpitts VCO is proposed. The good phase noise of this architecture makes it very suitable for high-end applications. Fig 5.22 shows the principle of building QVCOs. The system-level schematic of the QVCO consists of two differential common collector Colpitts VCOs, and two coupling sub-circuits (CN), which use trans-conductance amplifiers to couple the two VCOs. As explained in [San05], the VCOs must be cross coupled to generate quadrature signals and avoid in-phase oscillation. As a result of cross coupling the two VCOs, the circuit generates the four different phases 0°, 90°, 180° and 270°.

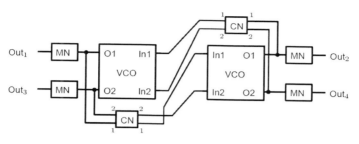

Fig 5.22 System level schematic: the circuit consists of two VCOs which are coupled via two coupling networks (CN), and four matching networks (MN) to 50 Ω

The device-level schematic of the common collector Colpitts QVCO is shown in Fig 5.23 and was designed in cooperation with Ms. Krause, [Bar10B]. Because the coupling between the two VCOs is done at the outputs of the VCOs and not at the tank, the two VCOs can be designed and optimized independently form the coupling network at the QVCO system-level: the performance of the QVCO will be as good as that of the two VCOs.

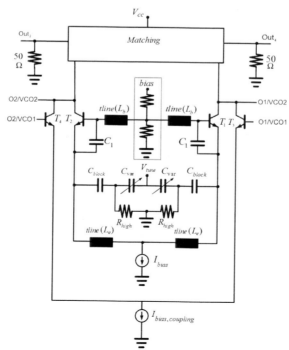

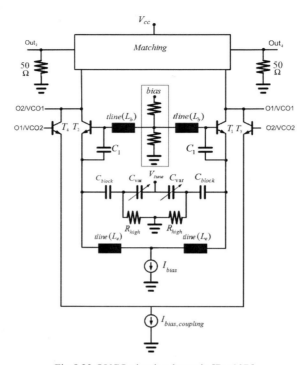

Fig 5.23 QVCO circuit schematic [Bar10B]

As a result, the design approach began with designing an optimized differential common collector Colpitts VCO. This was done by following the steps explained above. After that, the coupling circuit was designed. The QVCO was optimised for different transistor sizes and currents of the coupling circuit. Coupling the two VCOs at the output nodes and not in the tank was beneficial for two important reasons. First, the phase noise of the QVCO is independent to a high degree from the coupling circuit and is determined by the VCO phase noise. This was confirmed by simulations. Second, the oscillation frequency of this QVCO is independent from the coupling network current, whereas in other topologies [Ell09] it is. This can be explained by noting that coupling the current into the VCO tank directly affects the phase of the total current and hence, tunes the VCO to a new frequency (Barkhausen's criteria). As a result, variations in the magnitude of the coupling current affect the oscillation frequency.

In the final step of the design procedure, the load was optimized to present the optimal negative resistance to the tank of the VCO. That was done by loading the circuit with a parameterized load consisting of a series L and R circuit. Then, the load was swept for different values and the input impedance at one of the VCOs was simulated. Then, the load value which satisfied (5.36) was chosen. A linear decrease in the large signal trans conductance of the VCO with the VCO amplitude was assumed, which made (5.36) optimum for the choice of maximum power delivered to the tank, as given in [Bar10A]. The transmission line losses were estimated to be 4 Ω [Bar10A]. As a result, a negative resistance of -12 Ω is the optimum for the worst case negative resistance. After that, a matching network was designed to transform the 50 Ω load impedance into this value of the load over the tuning bandwidth. The optimization method is similar to what was shown in [Bar10A], but was improved by taking into account the loading effects of the coupling networks.

Chapter 6 Frequency Dividers

6.1 Introduction

As was mentioned in Chapter 4, frequency dividers are used in PLLs to divide the output frequency of the VCO into a lower frequency, which is suitable to be compared with the crystal oscillator frequency in the PFD. In the design of the 61.44 GHz PLL for the EASY-A project, the division ratio was chosen to be 1024, which was a result of choosing a crystal frequency of 60 MHz. This division ratio can be achieved by cascading ten divide-by-two circuits. As a result, in this thesis, only the design of divide-by-two circuits will be considered.

This chapter begins by explaining some of the characteristics of dividers. Then, static and dynamic frequency dividers are presented. After that, different latch architectures which can be used in building static dividers are introduced. Finally, the static-divider architectures are compared and the most suitable architecture is chosen for the design.

6.2 Divider Characteristics

In the design of a divide-by-two divider, three different characteristics of the divider have to be considered. In this section, those characteristics are explained.

6.2.1 Self-Oscillation Frequency

The self-oscillation frequency is the frequency at which the divider oscillates when no input is applied to the circuit. At this frequency, the divider needs minimum power at its input to operate in a proper way. This fact can be used to design the divider to oscillate at half the value of the frequency of interest. This is also beneficial in designing low power dividers, as will be explained later.

6.2.2 Divider Output Power

In designing cascaded frequency dividers for larger division ratios, the design by targeting a specific self-oscillation frequency and a specified output power leads to low-power designs. Therefore, the determination of the output power of a divider is necessary. The bias current, transistor sizes, and the load can be scaled to optimize for the self-oscillation frequency and output power of each divider stage. By doing this, each stage requires minimum input power and the output power of the previous stage can be scaled down, which finally results in lower power consumption.

6.2.3 Divider Sensitivity Curve

Depending on the frequency to be divided, the divider requires a minimum amount of power to function properly. To fully characterize the behavior of the divider at different frequencies, a plot, which relates the frequency and the minimum amount of input power for proper division at that frequency, is generated. Fig 6.1 shows a typical sensitivity curve of a divider. The explanation of the different regions can be found in [Sin02], as follows:

Chapter 6. Frequency Dividers

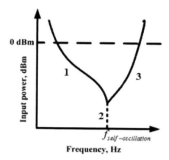

Fig 6.1 Typical divider sensitivity curve

- Region 1: when the frequency is low, low input power can still be enough for the divider to divide properly. The main reason is that the latching pair has enough time to restore the signal to the full level. However, the minimum input power at low frequencies is limited by the slew rate of the input signal, because if the slew rate is reduced, the divider becomes susceptible to oscillate near the zero input differential voltage. As a result, the divider does not work properly even at low power level at low frequencies.
- Region 2: at a certain frequency the divider may have the ability to oscillate and hence, the power required by the divider to divide double that frequency can be very small.
- Region 3: as the frequency increases, the clock differential pair would need higher slew rate to be able to switch the current in a fast manner. This high slew rate, translates to a higher signal swing for sinusoidal signals, [Raz98].

6.3 Divider Types

Two different types of divide-by-two frequency dividers exist, namely, static and dynamic dividers. In this section, both divider types are considered.

6.3.1 Dynamic Dividers

As the frequencies of PLLs or frequency synthesizers increase, the design of high speed frequency dividers becomes necessary. Dynamic dividers are the solution for high frequencies which is close to the f_t of the technology. This is mainly because this kind of divider exploits the low pass nature of the circuit in its operation, [Raz98]. In addition, dynamic dividers are loaded only by buffers, whereas in static dividers the output is loaded by the buffer and the master latch, which slows down the circuit. The most common use of such dividers is as a first stage in the PLL's frequency divider to divide the high frequency into a frequency at which static dividers are able to operate, [Kna03]. The dynamic frequency divider is based on the principle of regenerative frequency division. Fig 6.2 shows the divider block diagram. The input signal is applied to a mixer, which is followed by a low-pass filter and an amplifier. The output signal is fed back to the second mixer input. When using an active mixer no separate amplifier is necessary. The frequency response of the mixer conversion gain shows a low-pass behavior. Therefore no additional filter is required and the regenerative divider consists only of an active double-balanced mixer. The limited bandwidth of dynamic dividers is a drawback but poses no problem in many applications which require divider operation only in a specific frequency range. The operation principle of the dynamic divider can be clarified by noting that a mixer generated two components $f_{out} + f_{in}$ and $f_{out} - f_{in}$. If the former is filtererd, whereas the latter is fed back to be regenerated, equation (6.1) can be written:

$$f_{out} - f_{in} = f_{out} \tag{6.1}$$

which means

$$f_{out} = f_{in}/2 \tag{6.2}$$

Unfortunately, such a working principle dictates two limits for the frequency of operation, a low frequency limit when both $f_{out} + f_{in}$ and $f_{out} - f_{in}$ components of the input frequency can pass through the circuit and a high frequency limit which happens when the input frequency is two times the low pass filter bandwidth, which leads to attenuating the $f_{out} - f_{in}$ also.

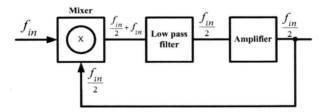

Fig 6.2 Regenerative divider [Kna03]

6.3.2 Static Dividers

A static frequency divider consists of two master-slave latches in a negative feedback configuration which forces the state of each latch to toggle between one and zero, [Raz98]. As a result of the feedback, the loading of the slave latch is approximately double that of the dynamic divider (the buffer and the master latch), which makes the speed of static dividers lower than their dynamic counterpart. Because the f_t of the technology is high enough to allow static dividers to work properly at 60 GHz, only static dividers are used in the design of the 1024 frequency divider. The master-slave divider architecture was adopted and is shown in Fig 6.3.

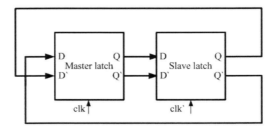

Fig 6.3 Master-slave static frequency dividers

6.4 Static Dividers Latch Architectures

In this section, four different latch topologies that can be used as the basic block of static frequency dividers are introduced.

6.4.1 Conventional Latch

A conventional latch, as shown in Fig 6.4, [Rog06], consists of two parts: the sampling pair, and the latching pair. The latch operates as follows: when the clock is low, the sampling differential pair is active and the latching pair is deactivated. As a result, the latch output follows the input. When the clock is high, the sampling pair is deactivated and the latching pair is active. In this case the latch restores and holds the last value. This is the simplest latch architecture. Unfortunately, there is a limit on the speed of this structure due to the low pass filtering of the parasitic capacitances at the collector and the load resistance, which limits the speed of the divider based on such a latch. The maximum speed of a conventional divider, as given in [Lee05], is in the range of 1/3 to 1/4 of the technology f_t.

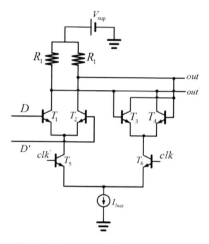

Fig 6.4 Conventional latch architecture

Because choosing faster latch architecture can lead to power saving, in the coming sections, three different speed enhancement techniques of the latch are discussed and the most suitable architecture for the divider is chosen.

6.4.2 Conventional Latch with Peaking

The performance and speed of a conventional latch can be improved by adding an inductor or transmission line in series with the load resistor. The role of this inductor is to introduce peaking in the latch bandwidth and extend its operation to higher frequencies. The conventional latch topology with peaking is shown in Fig 6.5, [Kna04]. Unfortunately, even though inductive peaking enhances the speed, the area of inductor or transmission line expands the circuit area. Additionally, the amount of peaking has to be carefully considered as it may lead to divider instabilities. As stated in [Lee98], when (6.3) is satisfied, a maximally flat frequency response is achieved with a factor of 1.72 bandwidth improvement of the latch compared to the conventional latch.

$$\frac{R^2C}{L} = 2.41 \qquad (6.3)$$

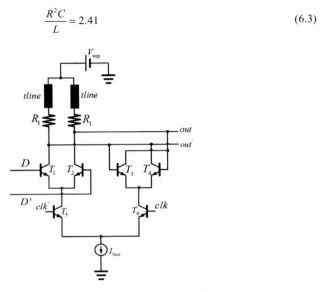

Fig 6.5 Conventional latch architecture with peaking

6.4.3 Split-Load Latch

The performance and speed of a conventional latch can also be improved by splitting the load of the latch into two loads R_1 and R_2 (R_1/R_2 is the split ratio); this makes it possible to control the gain of the sampling and latching pairs separately. The split-load latch topology is shown in Fig 6.6, as presented in [Lee05]. The advantage of this architecture is that it boosts the speed, and at the same time does not expand the area as is the case of inductive peaking. It has to be noticed that the ratio of the resistors should be carefully chosen, otherwise, it may lead to divider instability. As given in [Lee05], the improvement of this architecture can be understood by comparing the load of the latch with a cascade of RC circuits with values (R/n and C/n).

where R is the resistor of a conventional latch without splitting, C is the parasitic capacitance at the collector of the conventional latch and n is the number of segments that the load is split into.

The time constant of such a circuit can be given as:

$$\tau = \frac{(n+1)RC}{2n} \qquad (6.4)$$

Equation (6.4) predicts a reduction of 25% of the circuit time constant for two segments with a split ratio of unity. More improvement can be achieved for different split ratios, [Lee05].

Chapter 6. Frequency Dividers

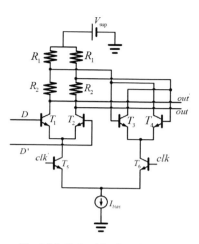

Fig 6.6 Split load latch architecture

6.4.4 Emitter Coupled Logic Latch

In order to enhance the speed of a conventional latch, emitter coupled logic (ECL) can be used, [Ryl03]. Fig 6.7 shows the ECL topology. In this topology, two emitter followers are used at the latch input and two emitter followers are used at the latch output. The role of these emitter followers is to amplify the current at the input and output of the latch, which results in faster charging and discharging of the capacitive nodes, leading to a faster latch. From another point of view, the capacitive loading in parallel with the load resistor is reduced due to buffering, leading to a reduced time constant on that node. Unfortunately, this architecture requires a higher supply voltage than the others.

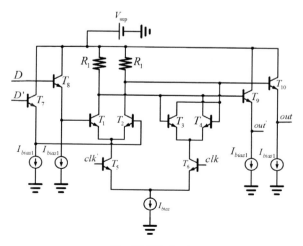

Fig 6.7 ECL latch

6.5 Simulations and Comparison

In order to compare the different latch architectures and to choose the most suitable one for the required divider, three different dividers were designed and optimized to give a 200 mV differential voltage swing on the output for the same self-oscillation frequency of 30 GHz, which is the optimum self-oscillation frequency for 60 GHz dividers. In order to optimize the divider, the following approach was followed:

The clock amplitude was set to zero and the data input was driven by a differential step function. After that, the delay of the latches was considered together with the loading effects. As stated in [Sea08], the self-oscillation of the divider can be calculated using the latch delay as given in (6.5).

$$f_{osc} = \frac{1}{2(\tau_{D1} + \tau_{D2})} \quad (6.5)$$

where τ_{D1} is the delay of the master latch and τ_{D2} is the slave latch delay. As a result, the latches were optimized to give a delay of around 16 ps, which, by using (6.5), corresponds to a 30 GHz self-oscillation frequency. Unfortunately, not every divider self oscillates. Therefore, in parallel to optimizing the latch, a transient simulation was run to guarantee that the divider oscillates at the target frequency.

The power consumed by the dividers was compared, keeping a constant power supply of 3 V for all three architectures. The results are shown in Table 6.1.

Table 6.1 Static divider architecture comparison (simulations)

Topology	Power Consumption (mW)
Conventional Latch	45
Conventional Latch with Peaking	25
Split Load Latch	30

As can be seen in Table 6.1, inductive peaking consumes the lowest power. Unfortunately, four inductors of approximately 1 nH each have to be added. This occupies much more area than a split-load divider. As a result, the split-load latch architecture was adopted. For the design of the divide-by-1024 frequency divider, the resistor ratio was controlled to change the latch delay and hence the self-oscillation frequency for each of the different stages in the divider. In addition, the power was scaled down for each stage. As given in [Kau06], equation (6.6) is a necessary condition for the divider to oscillate. This helps in scaling the latch resistors and choosing the right current of the transistors (transistor transconductance)

$$g_m R \geq 1 \quad (6.6)$$

where g_m is the latching pair transconductance and R is the load resistor.

Chapter 7 Phase Frequency Detectors and Charge Pumps (PFD/CP)

7.1 Introduction

The working principle of a PFD/CP is shown in Fig 7.1. The PFD compares the phases and frequencies of the reference signal and the output of the divider. Depending on the phases and frequencies of the inputs, the PFD generates up or down pulses. Those pulses command the charge pump to either inject charge to or draw charge from the loop filter.

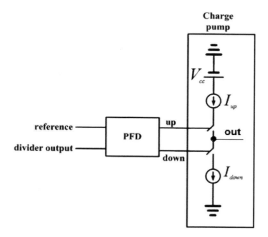

Fig 7.1 PFD/CP principle

This chapter begins by introducing the conventional phase frequency detector explaining its properties, such as, gain and dead-zone. After that, charge pumps are introduced, beginning with the architecture of a conventional charge pump. Then, improvements are added to the architecture to finally construct the highly matched charge pump architecture, which is superior to the conventional one with respect to reference feed through and voltage headroom requirements. Finally, the effect of the critical charge pump parameters on phase noise and reference feed-through is explained.

7.2 The Conventional PFD

Phase detectors produce an output signal proportional to the phase difference between its inputs. PFDs have the ability to also respond to frequency differences between the input signals. As a result, they are the most commonly employed phase detectors. The tri-state PFD, as shown in Fig 7.2, [Rog06], is the most common type of PFDs. In this PFD, if the reference clock leads the divided clock (its frequency is higher), the up signal is wider than the down signal. As a result, the charge-pump current raises the loop filter output voltage, hence, increasing the divided clock frequency until the frequency and phase of the reference clock and divided clock are aligned, and vice versa happens if the reference clock lags.

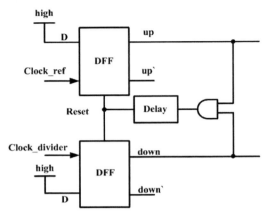

Fig 7.2 Tri-state PFD

7.3 PFD/CP Gain and Dead-zone

The transfer function of the PFD in the locked state is shown in Fig 7.3. As was mentioned in Chapter 4, the PFD transfer function can be described as:

$$K_{PFD} = \frac{I_{CP}}{2\pi} \tag{7.1}$$

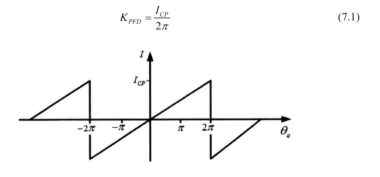

Fig 7.3 Tri-state PFD gain

A problem that occurs in this PFD is the dead-zone, as shown in Fig 7.4. As the phase difference between the inputs gets smaller, the pulses at the output of the PFD are not wide enough to switch the CP. Therefore, the gain becomes zero. In order to minimize the dead-zone, the delay shown in Fig 7.2 has to be used. By using this delay, the up and down signals have enough time to get to the full level and are able then to switch the CP currents which reduces the dead zone at small phase difference. The dead-zone degrades the close-in noise of the PLL. For the implemented PFD, the dead-zone was simulated to be 4° (200 ps at 60 MHz).

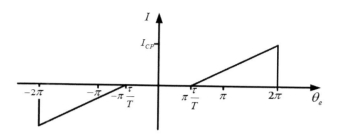

Fig 7.4 Dead-zone

7.4 Charge Pumps

The role of a charge pump in PLLs is to place charge into or take it out of the loop filter, hence, raising or reducing the output voltage of the loop filter, which controls the output frequency of the VCO. The input of the charge pump is connected to the output of the PFD, which commands the charge pump to inject or draw current from the loop filter according to the phase and frequency differences between the inputs of the PFD. Even though, several charge pump designs can be found in literature, the basic principle of a charge pump is the same, as shown in Fig 7.1.

As was mentioned in Chapter 4, the phase noise of a PLL is dominated by the noise of the charge pump. In addition, the charge pump is also responsible for reducing or increasing the spur power in the PLL's output spectrum. As a result, the design of the charge pump should be carefully considered and optimized for best performance.

7.5 Charge Pump Architectures

In this section, several charge pump architectures are considered.

7.5.1 Conventional Charge Pump

In the conventional charge pump, the switches in Fig 7.1 are implemented by using PMOS and NMOS transistors M_2 and M_3, respectively. The PMOS transistors are scaled three times the size of NMOS transistors for mobility reasons to improve symmetry. In addition, current mirrors are used to implement the current sources M_1 and M_4. Fig 7.5 shows the conventional charge pump. As will be explained later, the matching between the up and down currents of the charge pump is critical for the reduction of the spur power in the PLL spectrum. Unfortunately, due to channel modulation effects in transistors M_1 and M_4, the up and down currents depend on the value of the tuning voltage and cannot be matched for wide range of tuning voltages.

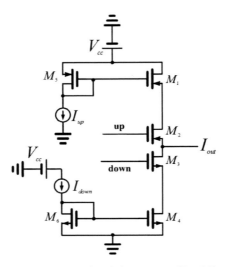

Fig 7.5 Conventional charge pump [Rog06]

As a result, the first improvement of the architecture is to improve the matching between the up and down currents. This matching is implemented in the highly matched charge pump architecture.

7.5.2 Highly Matched Charge Pump

In order to improve matching between the up and down currents an op-amp can be placed in the circuit, as shown in Fig 7.6. The op-amp implements a feedback, which improves the current matching between the up and down currents, [Rog06].

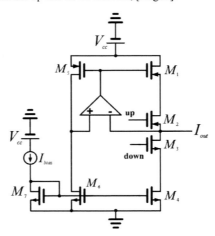

Fig 7.6 Highly matched charge pump [Rog06]

By using this feedback, the voltage on the drains of M_1 and M_4 is the same as the voltage at the drains of M_5 and M_6 which eliminates the channel modulation effect on the mismatch between both currents and allows an accurate copy of the current to be mirrored in both M_1 and M_4.

7.5.3 Highly Matched Charge Pump with Reduced Voltage Headroom

A draw-back of the architecture given in Fig 7.6 is the stacking of four transistors at the output. As a result, the required power supply to keep the current sources in saturation is increased. In order to reduce the voltage head room of the CP, the topology in Fig 7.7 can be used. By shifting M_2 and M_3 from the drains of M_1 and M_4 to their gates, the required voltage headroom is reduced. The current switching is now controlled using the gate voltages of M_4 and M_5.

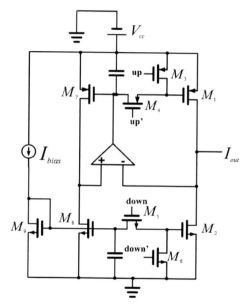

Fig 7.7 Highly matched charge pump with reduced voltage headroom [Rog06]

7.6 CP Design Issues

7.6.1 Charge Pump Noise

Because the noise of a charge pump can be the dominant source of noise in PLLs, it is important to consider the different sources of this noise and optimize the circuit in a way the minimizes their effect. In MOSFET transistors, two major noise sources at low frequencies are of concern, the channel noise and flicker noise, as shown in Fig 7.8.

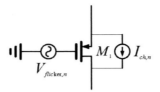

Fig 7.8 Major noise sources in MOSFETs

1. The channel noise depends on the transconductance of the transistors and is defined as, [Rog06]:

$$I^2_{ch,n}(f) = 4kT(\frac{2}{3})\sqrt{\mu C_{ox}\frac{W}{L}I_D} \qquad (7.2)$$

where k is Bolzmann's constant, T is the absolute temperature, μ is the carrier mobility, C_{ox} is the gate capacitance per unit area, W is the device length, L is the device length, and I_{DS} is the current through the device

2. Flicker noise: depends on the area of the transistor and other technology related parameters and is defined as, [Rog06]:

$$V^2_{flicker,n}(f) = \frac{K}{WLC_{ox}f} \qquad (7.3)$$

where K is a technology dependent parameter. This noise voltage can be transformed to a noise current on the output of the transistor by multiplying by the transconductance as:

$$I^2_{flicker,n}(f) = \frac{K\mu}{L^2 f}I_D \qquad (7.4)$$

Thus, the average total noise of two current sources operating only for a fraction t_{CP} of the period T_{ref}, can be calculated as:

$$I^2_{total}(f) = 2(\frac{K\mu}{L^2 f}I_D + 4kT(\frac{2}{3})\sqrt{\mu C_{ox}\frac{W}{L}I_D})\frac{t_{CP}}{T_{ref}} \qquad (7.5)$$

This result could be misinterpreted, that the current of the charge pump should be small for better noise performance. In order to understand the effect of charge pump noise on the phase noise of a PLL; this output current noise should be referred to the input by dividing by the PFD gain. The referred phase noise of the charge pump noise to the input of the PLL can be calculated as:

$$\varphi_n(f) = 2\pi\sqrt{2(\frac{K\mu}{L^2 I_D f} + 4kT(\frac{2}{3})\sqrt{\mu C_{ox}\frac{W}{L}})\frac{t_{CP}}{T_{ref}I_D^3}} \qquad (7.6)$$

As a result, the drain current should be large to minimize the effect of the charge pump noise on the PLL phase noise.

7.6.2 Current Sources Voltage Headroom

Because it limits the range of tuning voltages for which the charge pump works in an appropriately, it is important to consider the voltage headroom consumed by the current sources in the charge pump. The saturation voltage of NMOS can be calculated as:

$$V_{DS,sat} = V_{GS} - V_{th} = \sqrt{\frac{L}{\mu C_{ox} W} 2I_D} \qquad (7.7)$$

where V_{GS} is the gate to source voltage, and V_{th} is the threshold voltage of the transistor. As a result, the transistors' sizes in a charge pump tend to be large.

7.6.3 Current Sources Output Impedance

Due to the finite output impedance of MOSFET current sources and the different channel modulation effects in PMOS and NMOS transistors, it is important to keep the output impedance as high as possible and make the channel length modulation as low as possible. By doing so, the current mismatch is minimized. To achieve this, a longer gate length is beneficial. In addition, using source degeneration or cascode current sources increases the output resistance and improves the matching. Unfortunately, this consumes voltage headroom, and leads to higher supply voltages.

7.6.4 Charge Pump Reference Feed-through

Mismatches between the up and down currents in the charge pump causes spurs appearing in the output spectrum of the PLL at frequency offsets from the carrier which are multiple of the reference frequency. Assuming the up current is larger than the down current by ΔI, this can be explained by the following, [Rog06]:

1. At the locked state, the mismatch causes a charge of $q = \Delta I \cdot \Delta t_{res}$ to be injected into the loop filter, where Δt_{res} is the reset time of the PFD.

2. To remove this charge, the down current turns on for a time $\Delta t_{down} = \frac{q}{I_{down}}$.

3. This turns on the charge pump for a total time of $t_{CP} = \Delta t_{res}\left(1 + \frac{\Delta I}{I_{CP}}\right)$.

4. Assuming a conventional 2nd order loop filter, the charge is injected into the capacitor C_2. Therefore, the magnitude of the triangular wave form on the tuning voltage is $V_m = \frac{q}{C_2}$.

5. Due to the periodicity of this triangular waveform, its Fourier series can be calculated. The fundamental component of this Fourier series is:

$$V_{ref} = \frac{(\Delta t_{res})^2 \Delta I}{C_2 T_{ref}} \left(1 + \frac{\Delta I}{I_{CP}}\right) \quad (7.8)$$

where T_{ref} is the reference waveform period of oscillation.

6. For an approximation of the spurs' power, the fundamental component of the fourier series is used to modulate the VCO output, the result of this modulation is:

$$V_{out}(t) = A\cos(\omega_0 t) + \frac{AV_{ref}K_{VCO}}{2\omega_{ref}}\left(\cos(\omega_0 + \omega_{ref})t - \cos(\omega_0 - \omega_{ref})t\right) \quad (7.9)$$

where A is the amplitude of the output, ω_{ref} is the reference radian frequency, K_{VCO} is the VCO gain. Thus, the magnitude of the reference spurs relative to the carrier is:

$$Spurs = 20\log\left(\frac{(\Delta t_{res})^2 K_{VCO}\Delta I}{4\pi C_2}\left(1 + \frac{\Delta I}{I_{CP}}\right)\right) \text{ dBc}$$

$$(7.10)$$

For the implemented system, the delay was chosen to be Δt_{res} = 5 ns. Taking into consideration a mismatch current of 10% which corresponds to ΔI = 15 µA, K_{VCO} = 5 GHz/V, C_2 = 10 pF, and I_{CP} = 150 µA, the reference feed-through is predicted to be -35 dBc.

Chapter 8 Loop Filter

8.1 Introduction

In PLL systems, the phase detector is followed by a loop filter. The loop filter is necessary to filter out the high frequency contents of the output of the phase detector. In addition, it plays an important role in controlling the phase noise performance of a PLL and its dynamic behavior.

In designing the loop filter, the designer has to make several choices with respect to the filter order, the integration of the filter on-chip or including it on the PCB, and the implementation of an active or passive filter.

This chapter begins by introducing the different types of loop filters that can be used for a CP-based PLL. After that, the order of the loop filter is discussed. Finally, integrating loop filters using capacitor multipliers are presented.

8.2 Loop Filter Types

A loop filter can be implemented in different ways. In this section, the different types of loop filters, which can be interfaced with a charge pump, are introduced, as presented in [Rog06]. In addition, the trade-offs included in choosing one type or another are explained.

1. Passive RC: this type of filters uses only resistors and capacitors to implement the loop filter. Fig 8.1 shows the implementation of such a filter.

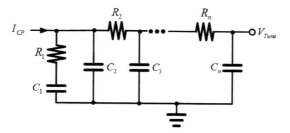

Fig 8.1 RC passive loop filter [Rog06]

The generic form of the filter transfer function is:

$$\underline{H}(s) = \frac{1+sT_1}{sC_t(1+sT_2) \times \prod_{i=3}^{n}(1+sT_i)} \quad (8.1)$$

where T_i are time constants. When the condition $T_i \geq 2T_{i+1}$ is satisfied, the time constants can be calculated as:

$$T_i = R_i C_i \quad \text{if } i \neq 2 \quad (8.2)$$
$$T_2 = R_1 C_1 (C_2 / C_t) \quad \text{if } i \neq 2$$

and $C_t = \sum_{i=1}^{n} C_i$.

2. Active RC: in this filter an operational amplifier (op-amp) is added to the passive RC structure, as shown in Fig 8.2. By doing so, the range of voltages of the charge pump can be adapted to the range of voltages required by the VCO. In addition, the design of the charge pump is relaxed, because the output voltage of the charge pump is centered at half the supply voltage and does not have to work rail to rail, where the MOSFET current sources are forced to operate in their triode region. Unfortunately, using an op-amp degrades the in-band noise performance of the PLL. As a result, active RC filters are avoided unless required.

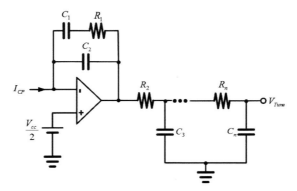

Fig 8.2 RC active loop filter [Rog06]

The transfer function of a third order active filter is given by:

$$H(s) = \frac{1+sR_1C_1}{s(C_1+C_2)(1+sC_sR_1)(1+sC_3R_3)} \quad (8.3)$$

where $C_s = C_1 \| C_2$. To maintain the fast settling time of a 2nd order loop, the higher order poles of the filter are chosen to be far from the dominant pole. As a result, they attenuate the higher harmonics of the charge pump output and do not affect the system's dynamic behavior. The relationship between the poles and zeros can be chosen as:

$$C_1R_1 = 10C_sR_1 = 10C_3R_3 \quad (8.4)$$

3. Active LC: in the passive and active RC filters, increasing the order of the filter can be only done by adding a single real pole to the filter, which corresponds to a 20 dB/dec roll-off in the transfer function. By using an LC filter, two benefits can be gained: first, the roll-off characteristics of the transfer function can be changed with complex poles. Second, the noise contributed by the resistors is eliminated. Unfortunately, LC filters cannot be integrated, due to the use of inductors. In order to adapt the well-known passive ladder LC filters to be used with charge pumps, a trans-impedance

amplifier at the input is added. An op-amp with an RC circuit in the feedback performs this conversion of current to voltage, as shown in Fig 8.3.

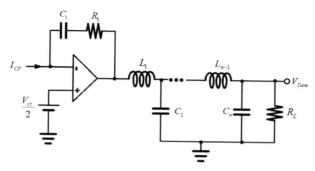

Fig 8.3 LC active loop filter [Rog06]

8.3 Loop Filter Order

The order of loop filter affects different loop parameters, such as, the attenuation of spur power, the stability of the loop, and the noise performance of the loop. Higher order loop filters should be carefully designed, because, otherwise, adding poles to the system could lead to instabilities.

8.4 Integrated versus Non-integrated Loop Filters

Loop filters can either be implemented off-chip, on-chip for a more compact solution. Using integrated filters eliminate the chance of using LC filters and limits the choices to either the passive or active RC filter. On the other side, using a non-integrated filter gives more control over the loop parameters and enables the use of sharper cut-off frequencies in the loop.

8.5 Capacitor Multiplier for Integrating Loop Filters

Because the cut-off frequency of a loop filter should be in the range of hundreds of kHz to several MHz, the capacitor values in the circuit can be large. As a consequence, the integration of capacitors on-chip requires large areas which can be expensive. To avoid that, capacitor multipliers can be used for that purpose. Unfortunately, even though using capacitor multiplier saves area, it adds noise to the circuit, as shown in Chapter 9. As a result, capacitor multipliers should be carefully used. In literature, two kinds of capacitor multipliers can be found. In this section, the different types of capacitor multipliers are explained.

8.5.1 The Conventional Capacitor Multiplier

The conventional capacitor multiplier, as presented in [Sol00], can be used to integrate the loop filter on chip. A current mirror can function as a capacitor multiplier, as shown in Fig 8.4. The structure uses a capacitor C_1 to generate a capacitor of $(N+1)C_1$, where N is the ratio between the two branches of the current mirror. A small signal equivalent circuit of the capacitor is shown in Fig 8.5. The analysis of the small signal circuits for the input admittance of the circuit is given in (8.5).

$$Y_{in} = \frac{1}{Z_{in}} = g_{o,MSN} + g_{o,MSP} + s(N+1)C_1 \frac{1 + \frac{sC_1}{(N+1)g_{MS_1}}}{1 + \frac{sC_1}{g_{MS_1}}} \quad (8.5)$$

where $g_{o,MSN}$ and $g_{o,MSP}$ are the output conductances of MSN and MSP, respectively, and g_{MS1} is the transconductance of MS1. Equation (8.5) shows two limits of the operation of the circuit as a capacitor:

- Low frequency limit coming from $g_{o,MSN} + g_{o,MSP}$, which limits the capacitor Q-factor.
- High frequency limit due to the pole coming from $\frac{sC_1}{g_{MS_1}}$.
- The zero can be neglected, because it is at a much higher frequency than the pole.

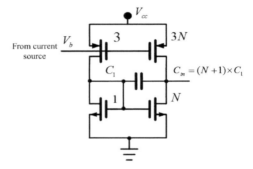

Fig 8.4 Conventional capacitor multiplier

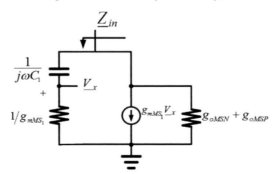

Fig 8.5 Small signal model of the conventional capacitor multiplier

8.5.2 The Self-biased Capacitor Multiplier

A capacitor multiplier can be implemented as shown in Fig 8.6, [Hwa06]. The benefit of this structure is that, it does not need any additional biasing circuitry no additional current sources are required. The input impedance of the circuit can be calculated as:

$$Z_{in} = \frac{\left(1+sC_1/\left(g_{m,MN_1}+g_{m,MP_1}\right)\right)/\left(g_{do,MN_2}+g_{do,MP_2}\right)}{1+SC_1\cdot(1+N)/\left(g_{do,MN_2}+g_{do,MP_2}\right)} \quad (8.6)$$

where g_{m,MN_1} and g_{m,MP_1} are the transconductances of MN_1 and MP_1, respectively, and g_{do,MN_2}, g_{do,MP_2} are the output conductances of MN_2 and MP_2, respectively. Equation (8.6) shows two limits of the validity of the circuit as a capacitor multiplier. First, a low-frequency limit, which is caused by the pole at:

$$p1 = \left(g_{do,MN_2}+g_{do,MP_2}\right)/\left(C_1(1+N)\right) \quad (8.7)$$

Second, a high-frequency limit, which is caused by the zero at:

$$z1 = \left(g_{m,MN_1}+g_{m,MP_1}\right)/C_1 \quad (8.8)$$

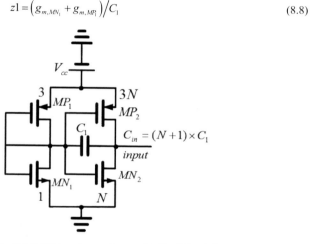

Fig 8.6 Self-biased capacitor multiplier [Hwa06]

In the range between the pole and the zero the circuit behaves as a capacitance. The value of this capacitance is:

$$C_{in} = (N+1)C_1 \quad (8.9)$$

In order to extend the usable bandwidth of the capacitor multiplier, the output impedance of the transistors MN_1 and MP_1 should be increased. This can be done by increasing the channel length of the transistors. In addition, using a cascode structure instead of single transistors boosts the output impedance even more. Fig 8.7 shows the architecture of the adopted self-biased capacitor multiplier.

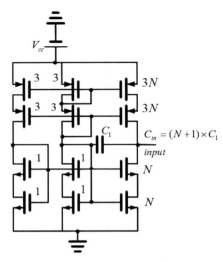

Fig 8.7 Cascode self-biased capacitor multiplier [Hwa06]

Chapter 9 Physical Design, Simulation and Measurement Results

9.1 Introduction

In this chapter, the simulation and measurement results of a common collector Colpitts QVCO, individual components of the PLL, the PLL, and the complete 60 GHz system (transmitter and receiver) are introduced. Regarding the PLL design, two versions were designed and measured. Version 1 is fully integrated and uses a capacitor multiplier for the loop filter integration. In version 2, to improve the phase noise performance, the CP was redesigned to allow more current and the loop filter was included off-chip to allow for the adjustment of the loop-bandwidth and the elimination of the capacitor multiplier. Furthermore, the output impedance matching to 50 Ω was improved. All components and chips were manufactured in IHP's BiCMOS SiGe HBT technology with an f_t of 180 GHz.

9.2 VCO

The common collector differential Colpitts VCO was measured on wafer. Due to the unavailability of balun probes at the time of measurement, one of the outputs was terminated on-chip using a 50 Ω resistor. The VCO chip-photo is shown in Fig 9.1. On the left side, a PGPGP probe was connected to supply the VCO with the bias current, supply voltage of 2 V, and tuning voltage in the range of 0 to 2.5 V. On the right side, a GSG probe to measure the RF output spectrum was connected. The spectrum of the VCO output is shown Fig 9.2. A phase noise performance of -98 dBc/Hz at 1 MHz frequency-offset was measured.

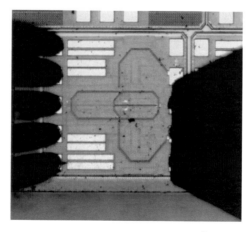

Fig 9.1 VCO photo 400 x 700 μm^2

For the phase noise and tuning range simulations, the simplified varactor model was used. The comparison of the simulation results and measurements results are shown in Fig 9.3 and Fig 9.4. As can be seen in Fig 9.3, a 15 % tuning range from 54 to 63 GHz was achieved. The tuning range measurement results are within 7% of the simulated values. The simple theory presented in Chapter 5 in section 5.7.2.1 could also be used for predicting the tuning range. The measurement of the varactor gave a capacitance range of 150 to 420 fF. This is equivalent to C_{equ} in the range of 56 to 131 fF when C_{block} = 300 fF and L_e = 159 pH are taken into account, when this value is substituted in (5.32) and the transistor model parameters in section 5.7.2.2 are used, the result is a frequency range of 49 to 62 GHz. In Fig 9.4, the phase noise measurements are compared to the simulations. It can be noticed that the measurement results are within 5 dB from simulations, which is a very reasonable result. The VCO consumes a power of 28 mW and the single-ended output power amounts to -5 ± 1 dBm. Detailed information about the design of the VCO can be found in [Bar10A].

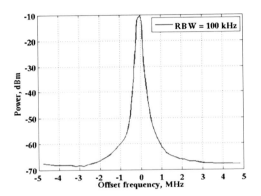

Fig 9.2 Measured VCO output spectrum at a frequency of 60.379 GHz, PN = -98 dBc/Hz @ 1 MHz offset frequency

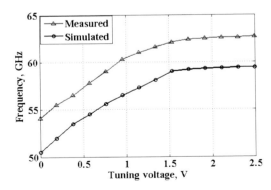

Fig 9.3 Measured and simulated oscillation frequency vs. tuning voltage

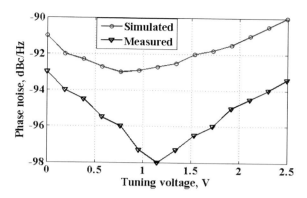

Fig 9.4 Measured and simulated phase noise vs. tuning voltage

9.3 Divide-by-1024 Frequency Divider

In order to split the design of the divide-by-1024 divider into blocks, a divide-by-two divider was designed and measured. Then, ten divide-by-two blocks were optimized using their self-oscillation frequency (depending on their input frequency) and the divide-by-1024 divider was implemented by cascading them. The frequency divider was measured on-wafer. Two balun probes, one for the input and the other for the output, were connected to the chip and a PGPGP probe was connected to supply the chip with the bias current, the supply voltage of 3 V, and the digital supply voltage of 2 V. The divider consumes a power of 90 mW. Fig 9.5 shows the measured self-oscillation frequency of the divider. The divider was designed and simulated to have a self-oscillation frequency of 60 GHz.

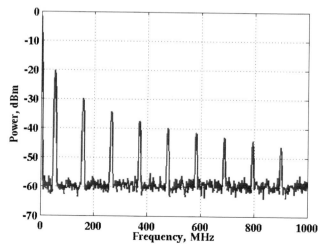

Fig 9.5 Measured divider self-oscillation frequency of 54 GHz = 52.7M x1024

The divider was driven by an input frequency of 61.44 GHz, and the output spectrum was measured. The result is shown in Fig 9.6.

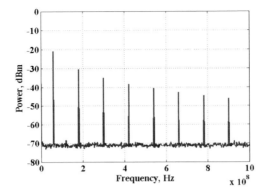

Fig 9.6 Measured divider output at 61.44 GHz input frequency, output frequency is 60 MHz

The sensitivity curve of the divider was also measured. The result is shown in Fig 9.7. As can be seen, the measured curve corresponds to the simulated one. There can be noticed a shift in the self-oscillation frequency due to parasitics.

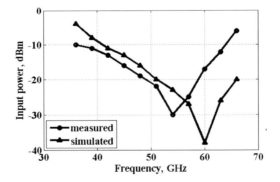

Fig 9.7 Measured and simulated divider sensitivity curves

9.4 Loop Filter

For the integrated PLL, a loop-bandwidth of 1.6 MHz was chosen, as stated in Chapter 4. The values C_1 = 110 pF, C_2 = 10 pF, R = 10 kΩ were chosen for the loop filter components. For the capacitor multiplier, to generate C_1 = 110 pF, a 10 pF capacitor was multiplied to 110 pF using a multiplication ratio of 11, as explained in section 8.5.2. Because for frequencies higher than 40 MHz the impedance of $C2$ is lower than 4 kΩ, $C2$ dominates the loop filter transfer function, due to the high resistor value of 10 kΩ. In addition, for frequencies higher than 400 kHz the impedance of $C2$ is lower than 3.6 kΩ. This makes R dominates the series combination of C_I and R. The previous discussion clarifies that the capacitor multiplier is only critical for frequencies lower than 400 kHz. Fig 9.8 shows the simulation result of the capacitance of the capacitor multiplier. For the loop filter, Fig 9.9 shows the simulation results of the comparison between the frequency responses of the integrated loop filter versus the conventional loop filter.

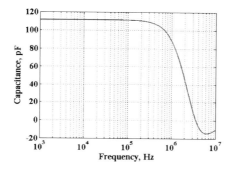

Fig 9.8 Simulated capacitance of the capacitor multiplier

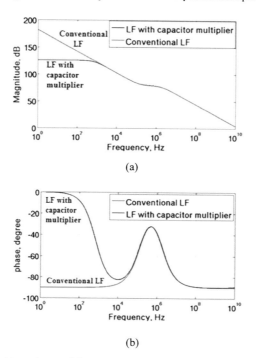

Fig 9.9 Simulated impedance of the conventional loop filter vs. loop filter using capacitor multiplier (a) magnitude response (b) phase response

Due to the low operation frequency of the loop filter, simulations were enough to show the suitability of using a capacitor multiplier for integrating the loop filter. The integrated loop filter was included in the system simulations to estimate the settling time of the PLL. Fig 9.10 shows the settling time of the PLL using the conventional and the integrated loop filters. A settling time of 4 µs was simulated using a loop-bandwidth of 1.6 MHz.

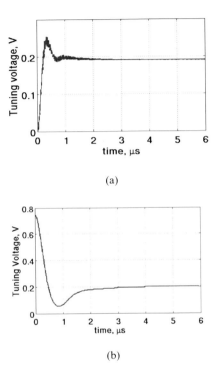

Fig 9.10 Simulated PLL dynamic behaviour (a) conventional LF (b) LF using the capacitor multiplier

9.5 Output Splitter

The output impedance of both the integrated-PLL and the improved non-integrated-PLL was measured using a balun probe. The measurement was done for two cases: The VCO is on, and the VCO is off. Fig 9.11 shows the measured output impedance of both versions of the PLL in both cases. As can be noticed, the matching at the oscillation frequency (61.44 GHz) was improved in version 2. It is also interesting to see the effect of oscillation on the magnitude of S_{11}, where a peak is observed (resonance-like peak) near the PLL output frequency. Because the mixer following the PLL includes a variable gain amplifier, a differential output power of -9 dBm was enough to drive the mixer. The output splitter was designed to deliver this amount of power to the mixer input. The output splitter draws a current of 8 mA from a 3 V power supply.

Chapter 9. Physical Design, Simulations, and Measurements Results

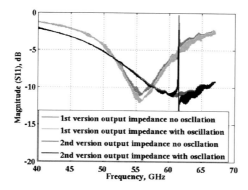

Fig 9.11 Measured magnitude of the output reflection coefficient

9.6 Integrated PLL

Fig 9.12 shows a photo of the fabricated PLL IC. For testing the closed loop system, a PCB was designed to generate the DC bias voltages and currents for the PLL, as shown in Fig 9.13; the chip was wire-bonded. The output spectrum was measured on-wafer using a balun probe. The result is shown in Fig 9.14. As can be seen in Fig 9.15 (a), the results match with the predicted results of the phase noise simulations (simulations include two parts, theory which derives the transfer functions of the different noise sources and a simulation part, which simulates the noise of the individual blocks in the PLL). In addition, simulations predict an improvement of around 5 dB in the phase noise performance when the capacitor multiplier is replaced by a real capacitor, as shown in Fig 9.15 (b).

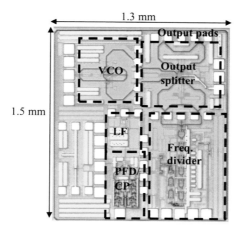

Fig 9.12 PLL chip-photo

The measured improvement of using a real capacitor instead of a capacitor multiplier is around 4 dB, as shown in Fig 9.16. It can be clearly noticed that, the measurement results match closely with the simulation results.

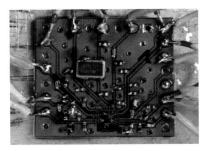

Fig 9.13 PLL PCB-photo

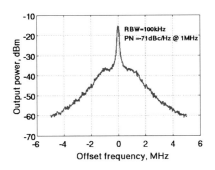

Fig 9.14 Integrated PLL output spectrum

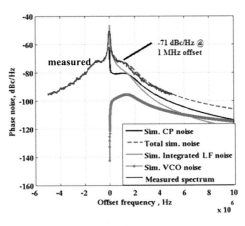

(a)

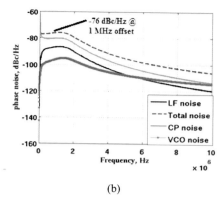

(b)

Fig 9.15 SSB phase noise (a) with capacitor multiplier measured versus simulated results (b) without capacitor multiplier simulated results

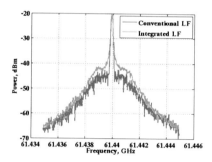

Fig 9.16 PLL measured output spectrum integrated vs. conventional (the effect of the integrated LF on phase noise)

9.7 Partly-integrated PLL

The performance of the integrated-PLL was sufficient for the purpose of the project; this is mainly due to using a carrier recovery circuit at the receiver-end. However, a new improved PLL, for applications where a carrier recovery circuit is not used, had to be designed. As a result, the phase noise performance of the PLL had to be improved. Therefore, the CP was redesigned to allow for higher currents to be injected into the loop filter. In addition, the loop filter was included off-chip to allow for the loop-bandwidth adjustment, which controls the loop dynamics versus the phase noise performance trade-off explained in Chapter 4 in section 4.9. The PLL was measured for three different loop-bandwidths and the results are shown in Fig 9.17 and Fig 9.18. As a result, the phase noise was improved by 12 dB (from -71 dBc/ Hz to -83 dBc/Hz at 1 MHz offset frequency) in comparison to the integrated-PLL (-4 dB gain from avoiding the capacitor multiplier and another -8 dB gain due to using a loop-bandwidth of 400 kHz, which is one fourth of the original loop-bandwidth. In addition, the power ratio between the carrier power and spurs power was improved by 12 dB (from -28 dB to -40 dB).

The effect of the charge pump current on the PLL spectrum was also studied. The spectrums for three different currents can be seen in Fig 9.19. The results show the three cases of the loop dynamics: over-damped, optimally-flat, and under-damped.

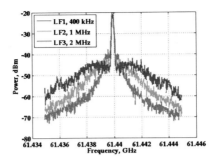

Fig 9.17 Measured output spectrum of the partly-integrated PLL for three different loop-bandwidths (RBW = 100 kHz)

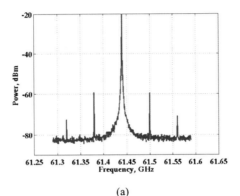

(a)

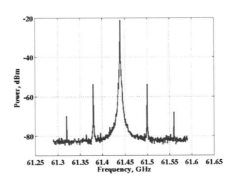

(b)

Chapter 9. Physical Design, Simulations, and Measurements Results   93

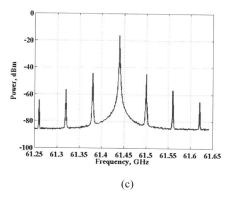

(c)

Fig 9.18 Spurs for three different loop-bandwidths (a) 400 kHz, -40 dBc (b) 1 MHz, -33 dBc (c) 2 MHz, -28 dBc

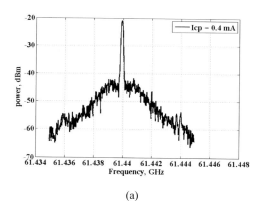

(a)

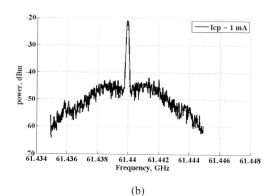

(b)

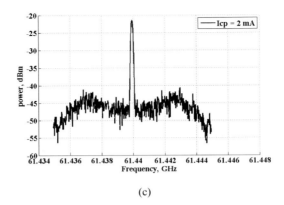

(c)

Fig 9.19 The effect of charge pump current on the output spectrum of the PLL (RBW = 100 kHz) (a) I_{cp} = 0.4 mA, over-damped (b) I_{cp} = 1 mA, optimally-flat (c) I_{cp} = 2 mA, under-damped

9.8 System Measurements

The PLL was used in a three chip solution transmitter, the modulator, the power amplifier, and the PLL. The three chips were placed inside a cavity. The role of the cavity is to make the bond-wires at the output of the power amplifier to the antenna as short as possible. As a result, the output power and the matching between the antenna and power amplifier are minimally affected. The antenna was included with its reflector on the PCB. The base-band signals were fed to the input of the modulator through SMA connectors and 100 Ω differential transmission lines. A separate raster-board to supply the PCB with bias currents and voltages was separately designed. This allows more flexibility in adjusting the currents and voltages of the PCB to achieve optimal performance. Fig 9.20 shows the PCB, the cavity, and the bonded three chips. The complete receiver and transmitter modules are shown in Fig 9.21 and Fig 9.22, respectively. The complete demonstrator measurement setup is shown in Fig 9.23.

Chapter 9. Physical Design, Simulations, and Measurements Results

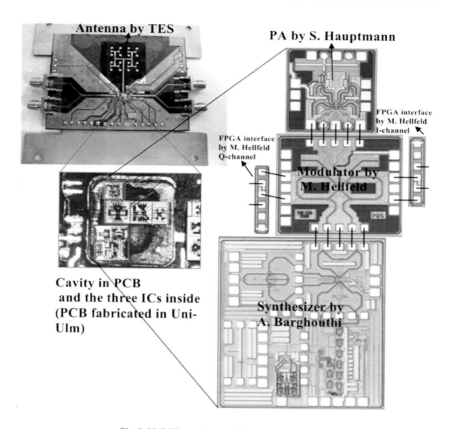

Fig 9.20 PCB, cavity and three chips solution

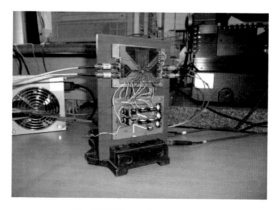

Fig 9.21 Transmitter module

Fig 9.22 Receiver module by Uni-Ulm

Fig 9.23 Demonstrator setup

The measurement setup consisted of a real time oscilloscope, power supplies, spectrum analyzer and the baseband modules. The measured recovered carrier is shown in Fig 9.24 and the measured eye diagram at the output of the receiver at a data rate of 3.65 Gbps using BPSK modulation is shown in Fig 9.25 for a distance of approximately 20 cm.

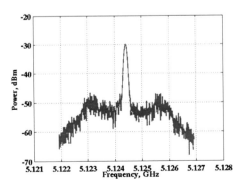

Fig 9.24 Recovered carrier (RBW = 100 kHz)

Chapter 9. Physical Design, Simulations, and Measurements Results

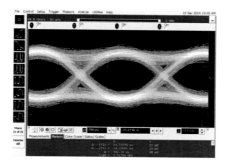

Fig 9.25 BPSK eye diagram at 3.65 Gbps

9.9 Common Collector Colpitts QVCO Simulations and Measurements

Designed in cooperation with Ms. Anna Krause [Kra09], the QVCO circuit was characterized on wafer. The measurement setup is shown in Fig 9.26. Two RF wafer-probes (GSGSG, left and right of the chip in Fig 9.26) were used for probing the four outputs, and one DC wafer-probe was used to supply the chip with the required biasing and the tuning voltage (PGPGP, top probe in Fig 9.26). Three of the outputs were terminated with 50 Ω and the fourth was connected to a spectrum analyser. The measured spectrum of the QVCO output is shown in Fig 9.27. The process was repeated for the four outputs. The tuning behaviour of the VCO is shown in Fig 9.28 as a function of the tuning voltage.

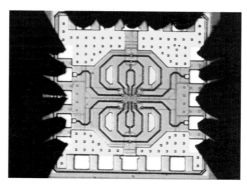

Fig 9.26 QVCO chip photo, designed by Ms. Anna Krause, [Kra09]

The QVCO can be tuned from 59.7 to 64.9 GHz, but taking the tuning bandwidth in the ± 1.5 dB change of the output power gives a tuning bandwidth from 59.7 to 62 GHz. The oscilloscope could have been used to measure the phases, but due to the high frequency of oscillation, the wave length of the signal in the cable is only around 3 mm, which means any cable bend or movement results in large measurement inaccuracies. The simulated output phases of the QVCO are shown in Fig 9.29. A summary of the measured QVCO performance is shown in Table 9.1. The simulated and measured phase noise performance is shown in Fig 9.30.

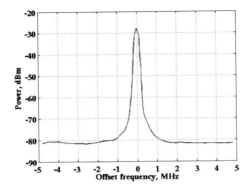

Fig 9.27 The measured QVCO spectrum at a frequency of 59.7 GHz (RBW = 100 kHz)

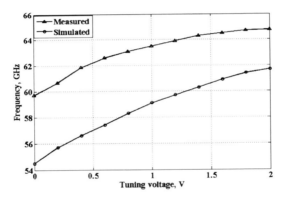

Fig 9.28 The measured and simulated tuning behaviour of the QVCO

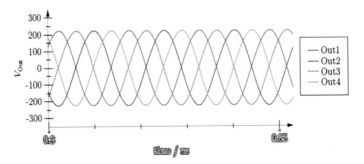

Fig 9.29 Simulated phases of the QVCO (0°, 90°, 180°, 270°)

Chapter 9. Physical Design, Simulations, and Measurements Results

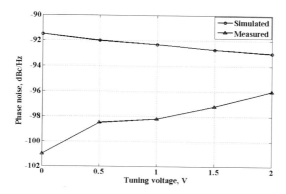

Fig 9.30 Simulated and measured phase noise performance

Table 9.1 Measured QVCO performance data

Technology	f_t =180 GHz
Tuning range	59.7 - 62 GHz
Output power	-13 ± 1.5 dB
Supply voltage	2 V
Phase noise	-101 to -96 dBc/Hz
Area	0.7 x 0.65 mm^2
Power consumption	64 mW

Chapter 10 Conclusion and Perspective

10.1 Introduction

In this work, the analysis and design of a 60 GHz QVCO and a 61.44 GHz PLL were introduced. In addition, new insight on the frequency limits of HBT cross coupled oscillators was presented, by including the effect of the HBT transistor parasitics into the analysis of the circuit negative conductance and deriving an expression for the maximum frequency, at which, the circuit conductance changes its sign from negative to positive. The HBT transistor parasitics were also taken into account into the analysis of the negative resistance of the differential common collector Colpitts oscillator and an extended expression for the negative resistance of the circuit was derived.

In this chapter, selected key results, in this thesis, are compared to state of the art components and designs. In addition, future works and improvements are indicated.

10.2 Comparison with State-of-the-Art

10.2.1 VCO

In Table 10.1, a comparison of the achieved results of the VCO with state-of-the-art VCOs is shown. To our knowledge, this is the best phase noise performance of a 60 GHz VCO in silicon yielding such a high tuning bandwidth.

Table 10.1 Comparison with state-of-the-art VCOs

Tech f_t/ GHz	Freq. Tuning/ GHz	Phase noise at 1MHz/ dBc/Hz	Output Power/ dBm	Supply power/ mW	Ref.
120	58.7- 68.5	-87 to -84	-17	73.8	[Win03]
150	75.3- 79.6	-95 to -93	14	-	[Li03B]
150	104 -107	-101	2.5	140	[Nic06]
180	54.1-62.7	-98 to -90	-6 to -4	28	**This work**

10.2.2 PLL

In Table 10.2, a comparison of the designed PLLs with state-of-the-art PLLs is shown. For a fair comparison between different designs, the loop noise can be normalized by the division ratio (N), as this linearly amplifies the noise to the output. With this assumption, a new figure of merit (FOM) can be defined by referring all PLLs to a loop bandwidth of 1 MHz assuming roll-off behavior of -20 dB/decade for the loop; this is an upper boundary performance, because -20 dB/decade is the maximum slop possible. The FOM allows the comparison of different loops with different bandwidths. The FOM is defined as:

Chapter 10. Conclusion and Perspective

$$FOM = PN - 20\log N + 20\log \frac{f_{offset}}{f_{BW}}, \quad f_{offset} > f_{BW} \quad (10.1)$$

$$FOM = PN - 20\log N \quad , \quad f_{offset} \leq f_{BW} \quad (10.2)$$

where f_{offset} is the offset frequency at which phase noise is calculated, PN is the phase noise at the given offset frequency and f_{BW} is the loop bandwidth.

Table 10.2 Comparison with state-of-the-art PLLs

Technology	f /GHz	PN at 1 MHz dBc/Hz	Loop B.W. / kHz	Ref. freq. / MHz	DC power / mW	FOM/ at 1 MHz dBc/Hz	Ref.
0.25μm BiCMOS	65 (fix.)	-82	200	60	650	-128	[Win05]
130nm CMOS	46 (fix.)	-72	500	60	57	-123	[Cao07]
90nm CMOS	61 (fix.)	-80	900	60	78	-141	[Hos07]
0.25μm BiCMOS	50-53 (var.)	-81	1000	262	400	-127	[Lee08]
0.25μm BICMOS	61.44 (fix.)	-71	1600	60	200	-131	This work
0.25μm BICMOS	61.44 (fix.)	-83	400	60	200	-135	This work

10.2.3 QVCO

In Table 10.3, a comparison of the achieved results of the QVCO with state-of-the-art QVCOs is shown. To our knowledge, this is the best phase noise performance reported for a silicon 60 GHz QVCO.

Table 10.3 Comparison with state-of-the-art QVCOs

Tech ft/ GHz	Freq. Tuning/GHz	Phase noise at 1 MHz dBc/Hz	Output power/ per channel in dBm	Supply power/ mW	Ref.
n. a.	23 - 24.4	-94	n.a.	22	[San05]
n. a.	59-62.5	<-80 (at 2 MHz)	-19 ±1.5	84	[Ell09]
85	24.8 - 28.9	-84.2	-14.7	129	[Hac03]
180	59.7 - 62	-101 to -96	-13±1.5	64	This work

10.3 Theoretical Extensions

In addition to the achieved results on the hardware level, extensions to the available theory in literature of cross coupled oscillators and Colpitts oscillators were done.

- The negative resistance expression of the Colpitts oscillator was extended by taking the base-resistance R_b and base-collector capacitance C_{bc} into account in the circuit analysis. The new expression is shown in (5.34). This enables the designer to estimate the negative resistance more accurately at the beginning of the design process.

- The oscillation limits of the HBT cross coupled oscillator were also studied by taking the HBT transistor parasitics into account in the circuit analysis. The results are shown in equations (5.24), (5.27), (5.28) and (5.31). Such results, gives the designer the ability to test the suitability of using the cross coupled oscillator for the target frequency using a certain technology.

10.4 Outlook

In the future, several design improvements can be made:

- An integrated-PLL with an improved phase noise performance for higher-end applications can be implemented. This can be done by including large capacitors on chip, which means better phase noise at the cost of larger area. Another option is to implement novel capacitor multiplier architectures, which minimally affect the phase noise performance of the PLL.
- The phase noise can also be improved by choosing a larger reference frequency. This can be achieved by designing another PLL, which works as the reference of the 61.44 GHz PLL.
- The quality of the quadrature signals at the input of the modulator can also be improved. This can be done by implementing a PLL, which is based on the implemented QVCO. By doing so, phase shifters and poly-phase filters can be avoided and the quadrature signals can be generated more accurately.

Appendix A

A. Verilog-A PLL Codes

A.1 VCO Verilog-A Code

```
`include "constants.vams"
`include "disciplines.vams"
`include "discipline.h"
`include "constants.h"
module vco(vin, vout);
input  vin;
output vout;
electrical vin, vout;
parameter real    vco_amp = 250m from (0:inf);
parameter real    vco_cf = 60G  from (0:inf);
parameter real    vco_gain = 3G exclude 0.0;
parameter integer vco_ppc = 40  from [4:inf);
    real wc;              // center freq in rad/s
    real phase_lin;       // wc*time component of phase
    real phase_nonlin;    // the idt(k*f(t)) of phase
    integer num_cycles;   // number of cycles in linear phase component
    real inst_freq;       // instanteous frequency
    analog begin
@ ( initial_step ) begin
wc = `M_TWO_PI * vco_cf;
end
// linear portion is calculated so that it remains in the +/- 2`PI range
// This is to ensure its value doesn't get too large and cause rounding
// problems for calculation of the phase.
```

```
phase_lin = wc * $abstime;

num_cycles = phase_lin / `M_TWO_PI;

phase_lin = phase_lin - num_cycles * `M_TWO_PI;

phase_nonlin = `M_TWO_PI * vco_gain * idtmod ( V(vin), 0, 1000.0, 0.0);

V(vout) <+ vco_amp * sin (phase_lin + phase_nonlin);

// ensure that modulator output recalculated soon.

inst_freq = vco_cf + vco_gain * V(vin);

$bound_step (1/(vco_ppc * inst_freq));

end

endmodule
```

A.2 Divider Verilog-A Code

```
`include "constants.vams"

`include "disciplines.vams"

`include "constants.h"

`include "discipline.h"

 module DIV_va (in,out);

 input in; output out; electrical in, out;

parameter real vlo=0, vhi=2.5;

parameter integer ratio=25 from [2:inf);

parameter integer dir=1 from [-1:1] exclude 0;

//dir=1 for positive edge trigger

//dir=-1 for negative edge trigger

 parameter real tt=100p from (0:inf);

 parameter real td=10p from (0:inf);

 parameter real ttol=1p from (0:td/5);

 integer count,n;

 real dt;

 analog begin
```

```
@(cross (V(in)-(vhi+vlo)/2, dir, ttol)) begin
count = count+1;
  if (count >= ratio)
  count=0;
  n=(2*count >= ratio);
end
V(out) <+ transition(n? vhi: 0, td, tt);
end
endmodule
```

A.3 PFD Verilog-A Code

```
`include "constants.vams"
`include "disciplines.vams"
module PFD(down,up,vco_clk,ref_clk);
 input vco_clk,ref_clk;
 output down,up;
electrical vco_clk,ref_clk;
electrical down,up;
parameter real threshold=1 ;
integer state;
analog begin
@(cross(V(ref_clk)-threshold,1))
if(state<1) state=state+1;
@(cross(V(vco_clk)-threshold,1))
if(state>-1) state=state-1;
V(up)<+transition((state==1)?2:0,1p,1p);
V(down)<+transition((state==-1)?2:0,1p,1p);
end
endmodule
```

A.4 Charge-pump Verilog-A Code

```
`include "constants.vams"

`include "disciplines.vams"

module charge_pump(Iout,down,up);

input up;

input down;

output Iout;

parameter real Ip=25e-6;

parameter real threshold=1;

parameter real vmax=1.8;

parameter real vmin=0.2;

integer state;

electrical up,down,Iout;

analog begin

 if ((V(up)> threshold) && (V(down)< threshold))

 I(Iout)<+ -Ip;

 else if ((V(up)< threshold) && (V(down)> threshold))

I(Iout)<+ Ip;

else I(Iout)<+ 0;

end

endmodule
```

Appendix B

B. HBT Small Signal Model

The small-signal transistor model used to model bipolars in this thesis is a simplified version of the vertical bipolar intercompany model (VBIC) model. It is depicted in Fig B.1. The values of the small-signal elements can be calculated from `operating point parameters` and *model parameters*. The equations can be found in Table B.1.

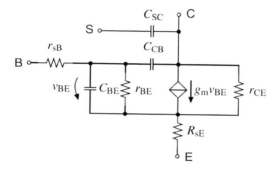

Fig B.1 Simplified HBT VBIC model

Table B.1 The calculation of the model parameters

parameter	description	equation
r_{sB}	base series resistance	rbi + rbx
R_{sE}	emitter series resistance	re
r_{BE}	base-emitter resistance	1/dib_dvbe
r_{CE}	collector-emitter resistance	-1/dic_dvbc
g_m	transconductance	dic_dvbe
C_{BE}	base-emitter capacitance	cje + cbex + *cbeo*
C_{CB}	collector-base capacitance	cbep + cjc + cbcx + *cbco*
C_{SC}	substrate-collector capacitance	cbcp

Appendix C

C. Matlab Codes

The matlab codes take into account the PLL transfer function and parameters, as given in Chapter 4.

C.1 Settling Time Simulations

```
function pll_step_response(pump_current,kvco,c1,c2,r,N)

pump_current=1e-3;

kvco=2*pi*5e9;

c1=2.2e-9;

c2=220e-12;

r=2e3;

N=1024;

t=0:0.5e-6:30e-6

num=[(pump_current*kvco*2*pi)/(2*3.14*c2) (pump_current*kvco)/(2*3.14*c2*r*c1)];

den=[8*pi*pi*pi (c1+c2)*(4*pi*pi)/(r*c1*c2) (pump_current*kvco*2*pi)/(2*3.14*c2*N)
    (pump_current*kvco)/(2*3.14*c2*N*c1*r)];

step(num,den,t)

axis([0 20e-6 0 2000]);

grid on
```

C.2 PLL Loop Bandwidth

```
function pll_loop_bandwidth(pump_current,kvco,c1,c2,r,N)

pump_current=1e-3;

kvco=2*pi*5e9;

c1=2.2e-9;

c2=220e-12;

r=2e3;

N=1024;

num=[(pump_current*kvco*2*pi)/(2*3.14*c2) (pump_current*kvco)/(2*3.14*c2*r*c1)];
```

den=[8*pi*pi*pi (c1+c2)*(4*pi*pi)/(r*c1*c2) (pump_current*kvco*2*pi)/(2*3.14*c2*N) (pump_current*kvco)/(2*3.14*c2*N*c1*r)];

bode (num,den)

grid on

References

[Agu04] Aguilera, J., Berenguer, R., *Design and Test of Integrated Inductors for RF Applications*, Kluwer Academic Publishers, New York, 2004, 1st edn.

[Bar10A] Barghouthi, A., and Ellinger, F.,: Design of a 54 to 63 GHz Differential Common Collector SiGe Colpitts VCO, *Proc. IEEE Conference on Microwave, Radar, and Wireless Communications*, MIKON, Vilnius, Lithuania, 2010, pp. 120-123

[Bar10B] Barghouthi, A., Krause, A., Carta, C., Ellinger, F.,Scheytt, C.,: Design and Characterization of a V-Band Quadrature VCO based on a Common-Collector SiGe Colpitts VCO, *IEEE CSICS*, 2010.

[Bar12A] Barghouthi, A., Joerges, U., Carta, C., and Ellinger, F.,: On the Oscillation Limits of HBT Cross Coupled Oscillators, *EuMA International Journal of Microwave and Wireless Technologies*, no. January 2012, 2012.

[Bar12B] Barghouthi, A., Hellfeld M., Carta, C., and Ellinger, F.,: Optimization of a 61.44 GHz BiCMOS HBT Integrated PLL for ultra-fast Settling Time, *EuMA International Journal of Microwave and Wireless Technologies*, no. January 2012, 2012.

[Beh07] Behzad, A.,: *Wireless LAN Radios System Definition to Transistor Design*, John Wiley & Sons, Inc., Hoboken, New Jersey, 2007, 1st edn.

[Bes03] Best, R.,: *Phase-locked Loops*, McGraw Hill, Oberwil, Switzerland, 2003, 4th edn.

[Cre08] Cressler, J.,: Fabrication of SiGe HBT BiCMOS Technology, Taylor & Francis Group, Bacon Raton, Florida, 2008, 1st edn.

[Cao07] Cao, C., Ding, Y., and Kenneth, K.: A 50-GHz Phase-Locked Loop in 130-nm CMOS, IEEE Journal of Solid-State Circuits, 2007, vol. 42, no. 8, pp. 21-24

[Che03] Cheng, D.,: *Field and Wave Electromagnetics*, Addison-Wesley Publishing Company, Canada, 1983.

[ECMA] Standard ECMA-387: High Rate 60 GHz PHY, MAC and HDMI PAL: http://www.ecma-international.org/publications/files/ECMA-ST/ECMA-387.pdf

[Ell07] Ellinger, F.,: *Radio Frequency Integrated Circuits*, Berlin Heidelberg, Springer, Germany, 2007, 2nd edn.

[Ell09] Ellinger, F., Jörges, U., and Hauptmann, S.,: Design and Analysis of a 59-62.5 GHz Quadrature LC Oscillator in 90 nm CMOS, *IET Journal on Circuits*, Vol 3, No. 6, Dec. 2009, pp. 322-330

[Gan12] Liu, G., et. al.,: A SiGe 7 Gbit/s Analog BPSK/QPSK Demodulator, *IEEE Conference for Microelectronics and Electronics (PRIME)*, Berlin, Germany, 2010, pp. 1-4

References

[Gra08]   Grass, E.,: D1.2_EASY_A_Requirements_SystemConcept_V1.0.doc

[Gut06]   Gutirrez, I., Melendez, J., et. al.: *Design and Characterization of Integrated Varactors for RF Applications*, John Wiley & Sons, Inc., West Sussex, England.

[Hac03]   S. Hackl, Boeck, J., et al.,: A 28 GHz Monolothic Integrated Quadrature Oscillator in SiGe Bipolar Technology, *IEEE Journal of Solid State Circuits,* Vol. 38, No,1, January 2003.

[Haj98]   Hajimiri, A. and Lee, T.,: A General Theory of Phase Noise in Electrical Oscillators. *IEEE Journal of Solid-State Circuits*, vol. 33, no. 2, February 1998, pp. 179-194

[Hau09]   Hauptmann, S., Ellinger, F., Korndoerfer, F., Scheytt, C.,: V-Band Variable Gain Amplifier applying Efficient Design Methodology with Scalable Transmission Lines, *IET Circuits, Devices and Systems*, May 2009.

[Her03]   Herzel, F., Winkler, W., and Borngräber, J.,: An Integrated 10 GHz Quadrature LC-VCO in SiGe:C BiCMOS Technology for Low Jitter Applications, *IEEE Custom Integrated Circuits Conference,* 2003.

[Hos07]   Hoshino, H., Tachibana, R., Mitomo, T., Ono, N., Yoshihara, Y., and Fujimoto, R.: A 60-GHz Phase-Locked Loop with Inductor-less Prescaler in 90-nm CMOS. *Proc. IEEE ESSCIRC Dig.*, Munich, Germany, 2007, pp. 472-475

[Hwa06]   Hwang, I.-C.,: Area-efficient and Self-biased Capacitor Multiplier for On-chip Loop Filter. IEE Electronics Letters, vol. 42, no. 24, August 2006, pp. 1392-1393

[Jeo06]   Joeng Y., Choi, S., Yang, K.,: Performance Improvements of InP-based Differential HBT VCO using the Resonant Tunneling Diode. *Proc. IEEE Conference on Indium Phosphide and Related Materials*, NJ, USA, 2006, pp. 42-45

[Kau06]   Kaushik S., Hashemi H.: Maximum Frequency of Operation of CMOS Static Freqeucny Dividers: Theory and Design Techniques, *Electronics, Circuits and Systems ICECS '06*, July 2006.

[Kna03]   Knapp, W., Wurzer, M., et. al.: 86 GHz Static and 110 GHz Dynamic Frequency Dividers in SiGe Bipolar Technology, *IEEE MTT-S*, 2003, pp. 1067-1070

[Kna04]   Knapp, H., Wurzer, M., 62-GHz 24-mW Static SiGe Frequency Divider, *IEEE SMIC*, 2004, pp. 5-8

[Kra09]   Krause, A.,: *Diploma Thesis*: Analysis and Implementation of a 60 GHz SiGe HBT Quadrature Voltage Controlled Oscillator, TUD-CCN, 2009.

[Kro11]   Krone, S.,: EASY-A_WG2_Final_Report.doc

[Kun06]   Kundert, K.: Predicting the Phase Noise and Jitter of PLL-Based Frequency Synthesizers, 2006, available: www.designers-guide.com

[Kur69]   Kurokawa, K.: Some Basic Characteristics of Broadband Negative Resistance Oscillator Circuits, *Bell System Technical Journal*, July 1969.

[Lee05]   Lee, T., Lee, H., et. al,: A 40-GHz Distributed-Load Static Frequency Divider, *IEEE Asian Solid-State Circuits Conference*, 2005, pp. 205-208

[Lee08]   Lee, J., Kim, H., and Yu, H.: A 52GHz Millimeter-Wave PLL Synthesizer for 60GHz WPAN Radio. Proc. *IEEE Asian Solid-State Circuits Conference*, Amsterdam, the Netherlands, 2008, pp. 155-158

[Li03A]   Li, H., Rein, H.-M, and Schwerd M.,: SiGe VCOs Operating up to 88 GHz suitable for Automotive Radar Sensors, *Electronic Letters*, vol. 39, no. 18, sept. 2003.

[Li03B]   Li, H., Rein, H.,: Millimeter-Wave VCOs with Wide Tuning Range and Low Phase Noise, Fully Integrated in a SiGe Bipolar Production Technology, *IEEE Journal of Solid-State Circuits*, vol. 33, no. 8, February 2003, pp. 184-190

[Man04]   Mansour Ma., Mansour, Mo., Mehorta A.,: Analysis of MOS Cross-Coupled LC-Tank Oscillators using Short-Channel Device Equations. *Proc. IEEE Design Automation Conference*, CA, USA, October 2004.

[Mar99]   Margarit, M., Tham, J.,: A Low-Noise, Low-Power VCO with Automatic Amplitude Control for Wireless Applications. *IEEE Journal of Solid-State Circuits*, vol. 34, no. 6, June 1999, pp. 761-771

[Nic06]   Nicolson, S., Rogers, K.H.K., et. al.: Design and Scaling of SiGe BiCMOS VCOs above 100 GHz, *IEEE BCTM*, 2006.

[Nic08]   Nicolson S., et. al.: A Low Voltage SiGe BiCMOS 77 GHz Automotive Radar Chipset, *IEEE Transactions on Microwave Theory and Techniques*, vol. 56, no. 5, May 2008

[Pet08]   Peter, M., Keugsen, W., Luo, J.,: Survey on 60 GHz Broadband Communication Capability, Applications and System Design, *Proc. IEEE European Microwave Integrated Circuits Conference*, Amesterdam, The Netherlands, October 2008, pp.1-4

[Raz98]   Razavi, B.,: *RF Microelectronics*, Prentice Hall Inc., Upper Saddler Rive, New Jeresy, 1998, 1st edn.

[Raz03]   Razavi, B.,: *Design of Integrated Circuits for Optical Communication*, Mc-Graw Hill, New York, USA, 2003.

[Rei05]   Noé, R.,: Phase Noise-Tolerant Synchronous QPSK/BPSK Baseband-Type Intradyne Receiver Concept Feedforward Carrier Recovery. *IEEE Journal of Lightwave Technology*, vol 23, 2005, pp.

[Rog03]   Rogers, J. and Plett, C.,: *Radio Frequency Integrated Circuit Design*, Norwood, MA, USA: Artech House, 2003, 1st edn.

References

[Rog06] Rogers, J., Plett, C., and Dai, F.: *Integrated Circuit Design for High Speed Frequency Synthesis*, Artech House, Norwood, MA, 2006, 1st edn.

[Ryl03] Rylayakov, A., Zwick, T.,: 96 GHz Static Frequency Divider in SiGe Bipolar Technology, *IEEE GaAs*, 2003, pp. 288-290

[Sad09] Sadri, A.,: IEEE802.15.3c Usage Model Document (UMD), Doc No: 15-06-0055-22-003c

[San05] Sanduleanu, M., and Stikvoort, E.,: Highly linear 24 GHz IQ oscillator, *IEEE Radio Frequency Integrated Circuits Syposium*, June 2005, pp. 577-578

[Sin02] Singh U., Green M.,: Dynamics of High Frequency CMOS Dividers, *IEEE International Symposium on Circuits and Systems ISCAS*, 2002.

[Sol00] Solis-Bustos, S., Silva-Martinez, J., et. al.: A 60-dB Dynamic-Range CMOS Sixth-Order 2.4-Hz Low-Pass Filter for Medical Applications. *IEEE Transactions on Circuits and Systems—II: Analog and Digital Processing*, Vol. 47, no. 12, Dec. 2000, pp. 1391-1398

[Ulu10] Ulusoy, A., et. al.: A SiGe Frequency Quadrupler for M-QAM Carrier Recovery. *Proceedings IEEE Topical Meeting on Silicon Monolithic Integrated Circuits in RFSystems (SiRF)*, New Orleans, LA, USA, Jan., 2010, pp.. 17 - 20

[Win03] Winkler, W., Borngräber, J., et. al.: 60 GHz and 76 GHz oscillators in 0.25 um SiGe:C BiCMOS, *IEEE ISSCC*, 2003.

[Win05] Winkler, W., Borngräber, J., Heinemann, B., and Herzel, F.: A Fully Integrated BiCMOS PLL for 60 GHz Wireless Applications. *Proc. IEEE International Solid-State Circuits Conference*, San Francisco, CA, USA, August 2005, pp. 406-407

[Yod06] Yodprasit U., Enz, C.C.,: Realization of a Low-Voltage and Low-Power Colpitts Quadrature Oscillator, *IEEE ISCAS*, 2006.

[Yod07] Yodprasit, U., Enz, C.C., and Gimmel, P.,: Common-mode Oscillation in Capacitive-coupled Differential Colpitts Oscillators, *Electronic Letters,* vol. 43, no. 21, Oct. 2007.

[Zou05] Zou, J., Mueller, D., et. al.: Fast Automatic Sizing of a Charge Pump Phase-Locked Loop based on Behavioral Models, *Proc. IEEE International Behavioral Modeling and Simulation Workshop,* 2005, pp. 100-105

Publications

[Bar10A] Barghouthi, A., and Ellinger, F.,: Design of a 54 to 63 GHz Differential Common Collector SiGe Colpitts VCO. *Proc. IEEE Conference on Microwave, Radar, and Wireless Communications,* MIKON, Vilnius, Lithuania, 2010, pp. 120-123

[Bar10B] Barghouthi, A., Krause, A., Carta, C., Ellinger, F.,Scheytt, C.,: Design and Characterization of a V-Band Quadrature VCO based on a Common-Collector SiGe Colpitts VCO, *IEEE CSICS*, 2010.

[Bar12A] Barghouthi, A., Joerges, U., Carta, C., and Ellinger, F.,: On the Oscillation Limits of HBT Cross Coupled Oscillators, *EuMA International Journal of Microwave and Wireless Technologies*, no. January 2012, 2012.

[Bar12B] Barghouthi, A., Hellfeld M., Carta, C., and Ellinger, F.,: Optimization of a 61.44 GHz BiCMOS HBT Integrated PLL for ultra-fast Settling Time, *EuMA International Journal of Microwave and Wireless Technologies*, no. January 2012, 2012.

[Jor11A] Joram, N., Barghouthi, A., Knochenhauer C., Ellinger, F., and Scheytt, C.,: Fully Integrated 50 Gbit/s Half-rate Linear Phase Detector in SiGe BiCMOS, in *Proceedings of the IEEE MTT-S International Microwave Symposium*, Baltimore, USA, 2011.

[Jor11B] Joram, N., Barghouthi, A., Knochenhauer C., Ellinger, F., and Scheytt, C.,: Logic Framework for High-Speed Serial Links in SiGe BiCMOS, *SCD Conference. Dresden*, Germany, 2011.

[Sch11] Schulte, B., Peter, M., Barghouthi, A., et. al.,: 60 GHz WLAN Applications and Implementation Aspects, *EUMA Journal*, 2011, pp. 1-9

Terms and Abbreviations

ADC	analog to digital converter
ADS	computer aided design program
AFE	analog front-end
AWGN	additive white Gaussian noise
BER	bit error rate
BiCMOS	bipolar complementary metal oxide semiconductor
CAD	computer aided design
CN	coupling network
CP	charge pump
DAC	digital to analog converter
DFT	discrete Fourier transform
EASY-A	enabled ambient systems – part A
ECL	emitter coupled logic
ECMA	wireless communication standard
EIRP	effective isotropic radiated power
EVM	error vector magnitude
FOM	figure of merit
f_t	transit frequency
f_{max}	maximum frequency of oscillation
FSK	frequency shift keying
GSG	ground-signal-ground probe
HBT	heterojunction bipolar transistor
HD	harmonic distortion
IDFT	inverse discrete Fourier transform
IF	intermediate frequency
IHP	semiconductor institute and foundary
IMD	intermodulation distortion
LF	loop filter
LNA	low noise amplifier
LO	local oscillator
LOS	line of sight
LTI	linear time invariant
LTV	linear time variant
Matlab	numerical programming language
MIM	metal insulator metal
NF	noise figure
NVA	network vector analyzer
OFDM	orthogonal frequency division multiplexing
PA	power amplifier
PCB	printed circuit board
PD	phase detector
PFD	phase frequency detector

PGPGP	power–ground-power-ground-power probe
PHY	physical layer
PLL	phase locked loop
PN	phase noise
Pnoise	periodic noise (spectreRF analysis)
PSS	periodic steady state (spectreRF analysis)
QAM	quadrature amplitude modulation
QPSK	quadrature phase shift keying
QVCO	quadrature voltage controlled oscillator
R_b	bit rate
RF	radio frequency
R_s	symbol rate
RTTT	road transport and traffic telematics
SC	single carrier
SNR	signal to noise ratio
SRF	self resonance frequency
STB	server for data processing
SiGe	silicon germanium
TR	tuning ratio
VCO	voltage controlled oscillator
UHR-C	ultra high rate cordless
VHR-E	very high rate extended range
Verilog-A	analog hardware description language

List of Figures

Fig 1.1 Trend of data rates in wireless communications [Pet08] 1
Fig 1.2 60 GHz frequency band [ECMA] 2
Fig 1.3 IEEE 802.15.3c / ECMA-387 channel plan [ECMA] 2
Fig 1.4 Allowed transmit power in different countries [ECMA] 3
Fig 1.5 Kiosk scenario file download (STB is a server) [Sad09] 4
Fig 1.6 UHR-C transceiver architecture [Kro11] 6
Fig 2.1 EVM calculation 9
Fig 2.2 Transmitter spectrum mask [ECMA] 10
Fig 2.3 I/Q mismatches [Raz98] (a) amplitude mismatch ($\theta=0$) 12
Fig 2.4 Simulated phase noise effect on the constellation diagram of an ideal communication link for different values at 1 MHz offset frequency (a) -71 dBc/Hz (b) -83 dBc/Hz (c) -90 dBc/Hz (d) -100 dBc/Hz 13
Fig 2.5 Superheterodyne transmitter 14
Fig 2.6 Direct conversion transmitter 15
Fig 2.7 Superheterodyne receiver 16
Fig 2.8 Direct conversion receiver 17
Fig 2.9 Transmitter Architecture 17
Fig 2.10 Adopted receiver architecture [Gan12] 18
Fig 3.1 Example of a SiGe HBT BiCMOS technology cross section [Cre08] 19
Fig 3.2 Resistor model 21
Fig 3.3 *CMIM* Model 21
Fig 3.4 L` and R` for microstrip line of 300 μm length, (+) EM simulations, (-) tline model [Hau09] 22
Fig 3.5 Inductive line model 22
Fig 3.6 Inductor model 23
Fig 4.1 PLL ideal spectrum 25
Fig 4.2 Phase noise non-ideality 25
Fig 4.3 Spur non-ideality 26
Fig 4.4 Tuning voltage transient 26
Fig 4.5 PLL block diagram 28
Fig 4.6 PLL linear model 28
Fig 4.7 The conventional and integrated loop filters 29
Fig 4.8 PLL noise model 30
Fig 4.9 Loop bandwidth versus a) phase noise b) settling time 33
Fig 5.1 The positive feedback model 35
Fig 5.2 The negative resistance model 36
Fig 5.3 Diode varactor [Gut06] 37
Fig 5.4 Inversion mode varactor [Gut06] 38
Fig 5.5 Accumulation mode varactor [Gut06] 39
Fig 5.6 Varactor test structure 39
Fig 5.7 Capacitance and Q-factor as a function of tuning voltage at 60 GHz (a) Capacitance (b) Q-factor 40

Fig 5.8 Accumulation mode varactor circuit model ... 41
Fig 5.9 Measurement setup [Gut06] ... 42
Fig 5.10 Setup parasitic [Gut06] ... 43
Fig 5.11 (a) Open (b) Single open, single short (c) Short, [Agu04] ... 44
Fig 5.12 The effect of a noise impulse on the output waveform of an oscillator [Haj98] ... 46
Fig 5.13 Cross coupled VCO ... 48
Fig 5.14 Cross coupled VCO small signal model ... 49
Fig 5.15 Simulated (all effects curve) and calculated parasitic effects (the other two curves) on the negative conductance of the cross-coupled oscillator using (5.21) and (5.24) ... 52
Fig 5.16 Simulated effect of C_m on the negative conductance of the cross coupled oscillator 52
Fig 5.17 Differential common collector Colpitts VCO [Bar10A] ... 53
Fig 5.18 A half circuit small signal model of the VCO ... 54
Fig 5.19 An equivalent half circuit of the emitter node at a specific frequency ... 55
Fig 5.20 Calculated and simulated effect of parasitics on the negative resistance of the VCO 56
Fig 5.21 Simulated negative resistance for different load values ... 58
Fig 5.22 System level schematic: the circuit consists of two VCOs which are coupled via two coupling networks (CN), and four matching networks (MN) to 50 Ω ... 59
Fig 5.23 QVCO circuit schematic [Bar10B] ... 60
Fig 6.1 Typical divider sensitivity curve ... 63
Fig 6.2 Regenerative divider, [Kna03] ... 64
Fig 6.3 Master-slave static frequency dividers ... 64
Fig 6.4 Conventional latch architecture ... 65
Fig 6.5 Conventional latch architecture with peaking ... 66
Fig 6.6 Split load latch architecture ... 67
Fig 6.7 ECL latch ... 67
Fig 7.1 PFD/CP principle ... 69
Fig 7.2 Tri-state PFD ... 70
Fig 7.3 Tri-state PFD gain ... 70
Fig 7.4 Dead-zone ... 71
Fig 7.5 Conventional charge pump, [Rog06] ... 72
Fig 7.6 Highly matched charge pump [Rog06] ... 72
Fig 7.7 Highly matched charge pump with reduced voltage headroom, [Rog06] ... 73
Fig 7.8 Major noise sources in MOSFETs ... 74
Fig 8.1 RC passive loop filter [Rog06] ... 77
Fig 8.2 RC active loop filter [Rog06] ... 78
Fig 8.3 LC active loop filter [Rog06] ... 79
Fig 8.4 Conventional capacitor multiplier ... 80
Fig 8.5 Small signal model of the conventional capacitor multiplier ... 80
Fig 8.6 Self-biased capacitor multiplier [Hwa06] ... 81
Fig 8.7 Cascode self-biased capacitor multiplier [Hwa06] ... 82
Fig 9.1 VCO photo 400 x 700 μm^2 ... 83
Fig 9.2 Measured VCO output spectrum at a frequency of 60.379 GHz, PN = -98 dBc/Hz @ 1 MHz offset frequency ... 84
Fig 9.3 Measured and simulated oscillation frequency vs. tuning voltage ... 84

List of Figures

Fig 9.4 Measured and simulated phase noise vs. tuning voltage 85
Fig 9.5 Measured divider self-oscillation frequency of 54 GHz = 52.7M x1024 85
Fig 9.6 Measured divider output at 61.44 GHz input frequency, output frequency is 60 MHz 86
Fig 9.7 Measured and simulated divider sensitivity curves 86
Fig 9.8 Simulated capacitance of the capacitor multiplier 87
Fig 9.9 Simulated impedance of the conventional loop filter vs. loop filter using capacitor multiplier (a) magnitude response (b) phase response 87
Fig 9.10 Simulated PLL dynamic behaviour (a) conventional LF (b) LF using the capacitor multiplier 88
Fig 9.11 Measured magnitude of the output reflection coefficient 89
Fig 9.12 PLL chip-photo 89
Fig 9.13 PLL PCB-photo 90
Fig 9.14 Integrated PLL output spectrum 90
Fig 9.15 SSB phase noise (a) with capacitor multiplier measured versus simulated results (b) without capacitor multiplier simulated results 91
Fig 9.16 PLL measured output spectrum integrated vs. conventional (the effect of the integrated LF on phase noise) 91
Fig 9.17 Measured output spectrum of the partly-integrated PLL for three different loop-bandwidths (RBW = 100 kHz) 92
Fig 9.18 Spurs for three different loop-bandwidths (a) 400 kHz, -40 dBc (b) 1 MHz, -33 dBc (c) 2 MHz, -28 dBc 93
Fig 9.19 The effect of charge pump current on the output spectrum of the PLL (RBW = 100 kHz) (a) I_{cp} = 0.4 mA, over-damped (b) I_{cp} = 1 mA, optimally-flat (c) I_{cp} = 2 mA, under-damped 94
Fig 9.20 PCB, cavity and three chips solution 95
Fig 9.21 Transmitter module 95
Fig 9.22 Receiver module by Uni-Ulm 96
Fig 9.23 Demonstrator setup 96
Fig 9.24 Recovered carrier (RBW = 100 kHz) 96
Fig 9.25 BPSK eye diagram at 3.65 Gbps 97
Fig 9.26 QVCO chip photo, designed by Ms. Anna Krause, [Kra09] 97
Fig 9.27 The measured QVCO spectrum at a frequency of 59.7 GHz (RBW = 100 kHz) 98
Fig 9.28 The measured and simulated tuning behaviour of the QVCO 98
Fig 9.29 Simulated phases of the QVCO (0°, 90°, 180°, 270°) 98
Fig 9.30 Simulated and measured phase noise performance 99

List of Tables

Table 1.1 UHR-C SC-QPSK PHY [Kro11] 6
Table 2.1 Cut-off frequencies of the spectrum mask [ECMA] 10
Table 4.1 Noise and loop transfer functions [Kun06] 31
Table 5.1 Model parameters analytical expressions, [Gut06] 41
Table 6.1 Static divider architecture comparison (simulations) 68
Table 9.1 Measured QVCO performance data 99
Table 10.1 Comparison with state-of-the-art VCOs 100
Table 10.2 Comparison with state-of-the-art PLLs 101
Table 10.3 Comparison with state-of-the-art QVCOs 101
Table B.1 The calculation of the model parameters 107

Curriculum Vitae

Personal Data

First Name	: Atheer
Family Name	: Barghouthi
Address	: Lautensackstr. 6
	80687 Munich
	Germany
Phone	: +49 (0)176 / 6518 4243
Email	: atheer.barghouthi@tu-dresden.de
Date and Place of Birth	: 24th.11. 1981, Ramallah, Palestine
Marital Status	: Married

Education and Academic Summary

- **2011-now** Senior design engineer for integrated analog optical transceivers in Silicon-Line company, Munich, Germany.

- **2008-2011** Scientific employee and Ph.D. candidate in the field of RF analogue circuit design in Dresden University of Technology. Thesis defense date: October, 2012.

- **2005-2008** Master of Science in Microelectronics & Microsystems in Hamburg University of Technology (graduated with distinction), Hamburg, Germany.

- **1999-2004** B.S.c Degree in Electrical Engineering (Major Computer & Communication) Grade: 85% (graduated with distinction), Birzeit University, Palestine.

- **1996-1999** General Secondary Examination (High School Certificate), Grade:95.6%, Ramallah, Palestine.

Work and Courses History

Work Experience:

- Currently, I am working in Silicon-Line as a senior design engineer in the design of integrated optical transceivers, Munichn, Germany.

- Four years working as a scientific employee in Dresden University of Technology, in that time, designed a 60 GHz PLL and additional RF circuits, Dresden, Germany

- Six months working as a mobile maintenance engineer in Jawwal Company, in Ramallah, Palestine.

Scholarships and Honours

- DAAD scholarship awardee to pursue my master degree. (2005 - 2007).
- Honors during my bachelor at Birzeit University, Palestine.
- Others for being an excellent student at school.